Teacher's Guide, Answer Key, and Supplementary Exercises

CCP#1

MATHEMATICS FOR CONSUMERS

by

Kathleen M. Harmeyer

and

Donald H. Jacobs

Media Materials, Inc.   •   Baltimore, Maryland

Editor
Barbara Pokrinchak, Ed.D.

Editorial Consultant
M. E. Criste

Published by Media Materials, Inc.
2936 Remington Avenue, Baltimore, Maryland 21211

Printed in the United States of America.

ISBN: 0-86601-071-8

CONTENTS

PREFACE

This *Teacher's Guide* to *Mathematics for Consumers* is meant to be a resource to the teacher in checking the students' work and in planning for further instruction.

Objectives: The basic objective of each topic has been included to help the teacher to focus on the main idea of the topic.

Vocabulary: Key words and difficult terminology have been indicated to help the teacher avoid vocabulary problems when the topic is taught.

Notes: Teaching tips and possible points of difficulty are indicated.

Answers: Answers to all exercises are included.

Activity Sheets: Worksheets have been included to provide additional practice in basic skills with whole numbers, decimals, and fractions. These worksheets may be duplicated for classroom use. Answers are given in the back of the Guide.

Chapter Tests: Two forms of a test for each chapter in the textbook have been included in this Guide. They may be used as a pretest and post-test for the chapter or as alternating forms of a summative test of the materials in the chapter. These chapter tests may also be duplicated for classroom use. Answers are provided for these, as well.

Chapter 1, EARNING MONEY (Pages 1-22)

Page 1, Wages

Objective: To compute weekly wages.

Vocabulary: Wages, hourly rate.

Notes: Discuss wages versus salary. Find out the current value of the minimum wage. Have students discuss what affects the amount of an hourly rate, such as difficulty of job, seniority, preparation of earner, responsibility factors, etc.

Encourage students to read the names of the people in the exercises.

ANSWERS:

1)	$200.00	6)	$239.07
2)	$173.28	7)	$368.80
3)	$300.00	8)	$382.40
4)	$358.00	9)	$137.20
5)	$391.09	10)	$314.80

Page 2, Annual Wages

Objective: To compute annual wages.

Vocabulary: Annual wages, estimate.

Notes: Discuss why 50 weeks is easier to calculate with than 52 weeks. Point out that many decisions are based on a person's annual income, e.g., how much money you can borrow from the bank, credit limits on credit cards, eligibility for assistance programs, etc.

ANSWERS:
1) $9,000
2) $7,500
3) $15,000
4) $7,000
5) $8,300
6) $10,500
7) $7,700
8) $13,500
9) $8,300
10) $11,300

Page 3, Time Worked

Objective: To compute elapsed time.

Vocabulary: A.M., P.M., elapsed time, renaming hours, 24-hour clock, hr., min.

Note: Do an example where students need to rename 60 minutes to one hour, such as #3.

ANSWERS:

	A.M.	P.M.	Total
1)	4	3	7
2)	3	4	7
3)	$4\frac{1}{2}$	$3\frac{1}{2}$	8
4)	5	4	9
5)	2	$4\frac{1}{2}$	$6\frac{1}{2}$
6)	$3\frac{1}{2}$	$5\frac{1}{2}$	9
7)	4	4	8
8)	4	5	9

Page 4:

1)	3 hr. 70 min.	2)	9 hr. 80 min.
3)	10 hr. 90 min.	4)	90 min.
5)	1 hr. 105 min.	6)	11 hr. 75 min.
7)	5 hr. 100 min.	8)	4 hr. 110 min.
9)	1 hr. 65 min.	10)	2 hr. 95 min.
11)	6 hr. 85 min.	12)	8 hr. 115 min.

Page 5:
1) 8 hours
2) 7 hours, 45 min.
3) 8 hours, 15 min.
4) 8 hours

Page 6:

1) Mon. 8 hr. 40 min.
   Tues. 8 hr. 15 min.
   Wed. 8 hr. 45 min.
   Thurs. 7 hr. 45 min.
   Fri. 8 hr.
   Total: 41 hr. 25 min.

2) Mon. 9 hr. 30 min.
   Tues. 9 hr. 30 min.
   Wed. 9 hr.
   Thurs. 9 hr.
   Fri. 9 hr.
   Total: 46 hours

Page 7, Overtime

Objective: To compute overtime wages.

Vocabulary: Time and a half, doubletime.

Notes: Point out that extra zeros in the product should be dropped. Hourly rates for time and a half are not rounded off. For example, in #7, $3.75 × 1.5 = $5.625, not $5.63.

ANSWERS:

Time and a Half
1) $6.00
2) $7.50
3) $5.94
4) $16.17
5) $9.75
6) $12.21
7) $5.625
8) $10.875

Doubletime
1) $8.00
2) $10.00
3) $7.92
4) $21.56
5) $13.00
6) $16.28
7) $7.50
8) $14.50

Page 8:

Notes: Do several examples of the problems with the students. This is a very challenging exercise. The very slow student will need assistance.

	Total Hours	Regular Hours	Overtime Hours	
			Time and a Half	Double Time
	a.	b.	c.	d.
1)	41	40	1	0
2)	40	40	0	0
3)	45	35	0	10
4)	44	40	1	3
5)	57	40	12	5
6)	46	40	2	4
7)	56	40	8	8
8)	43	40	0	3
9)	60	40	10	10
10)	56	40	8	8
11)	40	32	0	8
12)	43	32	3	8
13)	42	32	0	10
14)	42	32	2	8
15)	56	32	8	16
16)	41	32	1	8
17)	42	32	0	10
18)	46	32	6	8
19)	56	32	11	13
20)	46	32	0	14

Page 9, Regular Wages Plus Overtime

Objective: To compute total wages.

Notes: Encourage students to develop their own methods for solving these problems. The commutative property of multiplication provides a variety of methods.

ANSWERS:

	Regular Hours	Overtime Hours	Total Wages
1)	40	2	$193.50
2)	40	5	$180.50
3)	40	10	$451.00
4)	31	0	$207.70
5)	40	6	$490.00
6)	40	5	$246.05
7)	40	8	$507.52
8)	40	15	$558.75

Page 10:

1) $403.75	6) $390
2) $300	7) $226
3) $180	8) $306.18
4) $588.73	9) $322.92
5) $245	10) $269.50

Page 11, Tips

Objective: To compute earnings from tips.

Vocabulary: Tip (from "To Insure Promptness"), porter, skycap.

Note: Discusss occupations which depend on tips for income.

ANSWERS:
1) $350.50
2) $52.00
3) $250 for a 5-day week
4) $29.95
5) $320.00

Page 12, Piecework

Objective: To compute earnings from piecework.

Vocabulary: Piecework

ANSWERS:

1) $29.26	9) $12.48
2) $17.01	10) $23.14
3) $17.36	11) $31.52
4) $21.48	12) $29.19
5) $31.96	13) $29.55
6) $21.35	14) $30.60
7) $38.22	15) $51.03
8) $36.26	

Page 13:

	a.	b.
1)	42	$178.50
2)	46	$147.20
3)	41	$255.02
4)	29	$260.13
5)	50	$243.50
6)	104	$279.76
7)	200	$156.00
8)	347	$58.99
9)	294	$76.44
10)	450	$108.00

Page 14, The Key to Rounding Money

Objective: To round an answer to a specified place.

Vocabulary: Key digit

ANSWERS:

1) $5.02	2) $7.89	3) $10.49
4) $1.33	5) $30.00	6) $16.72
7) $3.78	8) $2.81	9) $86.12
10) $2.07	11) $91.56	12) $61.83
13) $2.94	14) $9.49	15) $22.46
16) $1.20	17) $3.48	18) $1.91
19) $1.24	20) $67.72	21) $2.07
22) $9.87	23) $78.65	24) $1.89
25) $8.26	26) $8.17	27) $5.36
28) $3.20	29) $5.37	30) $7.35

Page 15:

	Cent	Dime	Dollar
1)	$64.53	$64.50	$65
2)	$302.17	$302.20	$302
3)	$404.93	$404.90	$405
4)	$399.80	$399.80	$400
5)	$77.57	$77.60	$78
6)	$90.11	$90.10	$90
7)	$5.02	$5.00	$5
8)	$1.83	$1.80	$2
9)	$30.00	$30.00	$30
10)	$8.26	$8.30	$8

Page 16, Salary

Objective: To compute annual salary.

Vocabulary: Salary, semimonthly, bimonthly, quarterly, semiannually, salaried people.

Note: Discuss which types of jobs draw salaries.

ANSWERS:
1) $13,000
2) $13,200
3) $16,250
4) $18,000
5) $28,752
6) $57,024

Page 17:

	Times	Amount
1)	52	$230.77
2)	12	$1500.00
3)	4	$5000.00
4)	24	$666.67
5)	6	$2500.00
6)	26	$307.69
7)	2	$16,000.00
8)	24	$416.67
9)	26	$519.23
10)	52	$283.65
11)	12	$715.00
12)	52	$182.69

Page 18, The Key to Changing Percents to Decimals

Objective: To change percents to decimals.

Vocabulary: Percent

Note: Show examples of percent of a base other than 100. For instance, 400 squares with 20 squares shaded is 20 per 400, or 5 per 100.

ANSWERS:

1)	.06	2)	.07
3)	.035	4)	.094
5)	.32	6)	.01
7)	.015	8)	.125
9)	.10	10)	.50
11)	.666	12)	.75
13)	.15	14)	.062
15)	.0402	16)	.232

Page 19, Commission

Objective: To compute earnings from commissions.

Vocabulary: Commission, rate of commission.

Note: Point out that a *commission* is an amount of money and a *rate of commission* is a percent.

ANSWERS:

1)	$90.00	2)	$99.12
3)	$1040.00	4)	$810.00
5)	$3847.43	6)	$1676.29
7)	$344.12	8)	$128.95
9)	$111.10	10)	$2885.92

Page 20, Salary Plus Commission

Objective: To compute earnings from salary plus commission

Notes: When using the computer application, bring the computer into the classroom, is possible. Key in the program and test it before presenting it to the class.

Be certain students do not type in any punctuation marks except the decimal point.

To round off the total earnings in line 80 use:

$$E + INT(S * R + .5)/100$$

Have students design more problems for extra practice.

ANSWERS:

	Commission	Total Earnings
1)	$1200	$1700
2)	$3000	$3300
3)	$14,100	$14,350
4)	$44,285.01	$44,435.01
5)	$4913.19	$5113.19

SAMPLE RUN OF THE PROGRAM:

```
RUN
INPUT THE TOTAL SALES.
 120000
INPUT THE RATE OF COMMISSION
WITHOUT THE PERCENT SIGN.
 2.5
INPUT THE SALARY EARNED.
 300
TOTAL EARNINGS: $3300
```

Page 21, Net Pay

Objective: To compute net pay.

Vocabulary: Gross pay, net pay, take-home pay, deductions.

Note: Discuss other items which may be deducted. Collect sample pay check stubs to share with the students.

ANSWERS:

	Total Deductions	Net Pay
1)	$227.25	$372.75
2)	$152.92	$347.08
3)	$125.51	$277.49
4)	$25.49	$51.48
5)	$63.64	$57.11

Page 22, Chapter Review

ANSWERS:
1) .43   .138   .09   .875
2) $237.50
3) 8 hours
4) $255.59
5) $14.85
6) $25.20
7) $895.83
8) $355.77
9) $17,000
10) $148.03

Vocabulary Extras for Chapter 1

Draw these word pictures on the board and ask students which vocabulary words they suggest. Then encourage students to create their own word pictures.

A. Time Time	B. Time Ti	C. Paid / Time	D. RATE
E. $\mathscr{S}$	F. ANNUAL	G. Weekly Weekly	H. Per ¢

A. Doubletime   B. Time and a half
C. Paid overtime D. Piece rate
E. A dollar tip   F. Semiannual
G. Biweekly   H. Percent

CHAPTER 2, BUYING FOOD (Pages 23-38)

Page 23

Objective: To change prices from dollars to cents and from cents to dollars.

ANSWERS:
1) $.45   2) $.25   3) $.05
4) $.255   5) $1.39   6) $.36

1) 75¢   2) 80¢   3) 9¢
4) 345¢   5) 129.9¢   6) 489¢

Page 24, Reading Prices

Objective: To read prices as marked on grocery store items.

Notes: Bring in several items to demonstrate the various methods of marking prices.

ANSWERS:
1) 17¢ $.17   .17
2) $.99   99¢   .99
3) .68   68¢ $.68
4) .25 $.25   25¢
5) 89¢   .89 $.89

1) $1.09   1.09 1^{09} 1^{09}
2) 4^{39} $4.39   4.39 4^{39}
3) 3.99 3^{99} $3.99 3^{99}
4) 1^{69} 1.69 $1.69 1^{69}
5) 2^{98} 2^{98} 2.98 $2.98

Page 25, Adding Prices

Objective: To compute the total cost of food items.

Vocabulary: lb., oz., pkg.

Notes: For a practical laboratory activity, assemble a number of items from the grocery store. Set them out in groups and ask students to find the total cost of each group of items.

ANSWERS:
1) $4.18   2) $2.37
3) $4.87   4) $12.44

Page 26, Computing Change

Objective: To compute change.
Vocabulary: gal., doz.

Page 27

ANSWERS:

	Total Cost	Change		Total Cost	Change
1)	$6.35	$3.65	2)	$7.54	$2.46
3)	$9.94	$.06	4)	$6.06	$3.94
5)	$8.96	$1.04	6)	$3.37	$6.63
7)	$2.65	$7.35	8)	$3.67	$6.33
9)	$6.72	$3.28	10)	$5.47	$4.53
11)	$7.84	$2.16	12)	$9.04	$.96

Page 28, Store Coupons

Objective: To compute the cost of an item using a store coupon.

Notes: Display several examples of cents-off coupons. Discuss the differences between manufacturers' coupons and grocery store coupons. Do the exercise for a store which offers double value coupon redemption.

ANSWERS:

1)	$1.04	6)	$1.45
2)	$4.49	7)	$.86
3)	$1.02	8)	$.33
4)	$.42	9)	$.85
5)	$1.74	10)	$1.64

Page 29, Coupons for More Than One

Objective: To compute the cost of multiple items using a store coupon.

ANSWERS:

1)	$4.33	6)	$2.97
2)	$2.72	7)	$7.58
3)	$1.57	8)	$1.43
4)	$1.58	9)	$4.18
5)	$1.64	10)	$1.26

Page 30, Coupons with Conditions

Objective: To read information on coupons.

Vocabulary: Expire

ANSWERS:

Cruncho:
1) 40¢
2) Cruncho Chips
3) Two
4) 8 ounces
5) No
6) January 31, 1992

Real Mayonnaise:
1) 12¢
2) Real Mayonnaise
3) Two or one
4) 2 pints, one quart, or one 48-oz. jar
5) Yes
6) July 31, 1993

Page 31, Expiration Dates

Objective: To compute the number of months left before a coupon expires.

Vocabulary: Expiration

ANSWERS: May vary by $\frac{1}{2}$ month.

1)	10½ months	5)	$\frac{1}{2}$ month
2)	5 months	6)	3 months
3)	3 months	7)	4 months
4)	5 months	8)	6 months

Page 32, Finding the Cost of One

Objective: To compute the cost of one item.

Vocabulary: 2/99¢

Notes: This is different from computing the unit price in that when you compute cost per ounce, you rarely purchase exactly one ounce.

Page 33
ANSWERS:

	Cost of One Item	Total Cost of Single Items	Savings by Buying Items Together
1)	25¢	50¢	1¢
2)	50¢	$1.00	1¢
3)	20¢	$1.00	1¢
4)	43¢	$1.29	0¢
5)	13¢	91¢	2¢
6)	9¢	$1.08	8¢
7)	20¢	$1.00	0¢
8)	17¢	$1.02	3¢
9)	18¢	$1.44	5¢

Extra Exercises:

1. Choose an item from your pantry or cupboard at home and do this exercise with the price marked on the item.

2. Choose an item advertised in the newspaper and do this exercise with the price marked.

3. Does the multiple price always save you money? Explain.

Page 34, The Key to Using the Word "Per"

Objective: To replace the word "per" with "divided by."

Notes: The word "per" is used throughout the text and in many practical consumer ralated problems. The student should become alert to its use.

ANSWERS:

1) liters$\overline{)\text{cost}}$ 5) liters$\overline{)\text{kilometers}}$

2) hours$\overline{)\text{miles}}$ 6) kilograms$\overline{)\text{cost}}$

3) 100$\overline{)\,6.00}$ 7) seconds$\overline{)\text{meters}}$

4) minutes$\overline{)\text{feet}}$ 8) dozens$\overline{)\text{cost}}$

9) 100$\overline{)\,15.00}$

10) weeks$\overline{)\text{liters}}$

11) 100$\overline{)\,17.00}$

12) minutes$\overline{)\text{miles}}$

13) years$\overline{)\text{kilometers}}$

14) inches$\overline{)\text{cost}}$

15) games$\overline{)\text{attendance}}$

16) trips$\overline{)\text{miles}}$

17) days$\overline{)\text{newspapers}}$

18) seconds$\overline{)\text{inches}}$

19) pounds$\overline{)\text{cost}}$

20) weeks$\overline{)\text{gallons}}$

Page 35, The Unit Price

Objective: To compute unit prices.

Notes: Emphasize the use of the word *per*. Have students state each problem aloud, using the word, *per*; e.g., in #1, we are to find the cost *per* dozen.

ANSWERS:

1) $1.04   5) $.57
2) $.03   6) $.50
3) $.10   7) $.26
4) $1.92   8) $.11

Page 36, Comparing Unit Prices

Objective: To find the lowest unit price of a group of items.

Vocabulary: Brand names.

Notes: Collect various brands of the same item. Bring them to class and have students compare the prices. Do a taste test at the same time. Discuss what other factors affect a consumer's decision to buy.

Notes: Point out how small .20¢ is (a twentieth of one cent). Compare with the size of one gram. Distinguish between $.20 and .20¢. Help students interpret the answers to their divisions.

ANSWERS:

	A	B	C	D	Best Buy
1)	.60¢	.72¢	.32¢		C
2)	44¢	65¢	60¢	$1.22	A
3)	.48¢	.62¢	.44¢	.45¢	C
4)	$5.45	$4.49	$3.90	$4.42	C

Page 38, Chapter Review

1) $.89   2) $.07   3) $1.69
4) 13¢
5) The best buy is Saltridge Farms Mix at $.12/oz. The change is $4.17.

6)

Size	Jars Needed	Cost per jar	Total Cost	Cost with Coupon	Total Ounces	Cost per Ounce
12	2	1.29	2.58	2.18	24	9¢
18	2	1.69	3.38	2.98	36	8¢
28	1	1.99	1.99	1.59	28	6¢
48	1	2.49	2.49	2.09	48	(4¢)

Best Buy

Vocabulary:
What words do these pictures suggest?

A.   B.   C.   D.

A. Cents-off Coupon
B. Unit Price
C. Multiple price
D. Compare (COM pair)

CHAPTER 3, SHOPPING FOR CLOTHES (Pages 39-58)

Page 40, Ready-to-Wear

Objective: To compute the cost of clothing purchases, including tax.

Vocabulary: Sales tax, @.

Notes: Some sales taxes are rounded mathematically, i.e., to the nearer cent. Others are rounded to the next cent and are known as "bracket" taxes. Ascertain the method for computing sales tax, if any, in your locale. Discuss the financial benefits to the tax assessors of the bracket type taxes.

Page 41

ANSWERS:
1)	$41.94	6)	$83.99
2)	$40.77	7)	$86.73
3)	$26.80	8)	$176.99
4)	$48.08	9)	$80.98
5)	$84.32	10)	$422.64

Page 42, Sale Prices

Objective: To compute the amount saved on sale items.

ANSWERS:
1)	$7.01	6)	$.50
2)	$1.51	7)	$8.95
3)	$29.10	8)	$5.35
4)	$5.85	9)	$5.11
5)	$2.41	10)	$70.02

Page 43, Percent Saved

Objective: To compute the percent saved on sale items.

Notes: Not all calculators use the same logic. The calculator description does not apply to calculators designed with Reverse Polish Notation.

Estimation can be introduced here. Round $5.91 to $6.00 and $45.90 to $46.00. Apply the rule:
$$\frac{6}{46} = 13\%$$

ANSWERS:
1)	14%	6)	39%
2)	20%	7)	25%
3)	39%	8)	36%
4)	37%	9)	24%
5)	24%	10)	47%

Page 44, Discounts

Objective: To compute the sale price, using a discount.

Vocabulary: Discount, regular price.

Note: Students may develop alternate algorithms to solve these problems. Encourage this.

ANSWERS:
	Sale Price	Amount Saved
1)	$17.99	$2.00
2)	$29.74	$5.25
3)	$13.49	$1.50
4)	$31.96	$7.99
5)	$18.74	$6.25
6)	$110.49	$19.50
7)	$57.60	$6.40
8)	$75.00	$75.00
9)	$15.83	$2.16
10)	$17.99	$12.00

Page 45, Buying from a Catalog

Objective: To read catalog descriptions and to be able to order from a catalog.

Vocabulary: Catalog, 16 oz. = 1 pound. Some abbreviations used in catalogs may be new to the students.

Notes: Bring in examples of catalogs; discuss benefits and drawbacks of shopping from catalogs.

Have students determine their own sizes.

ANSWERS:
1)	$47.70	2)	$15.90

Page 46

1) 34 oz. = 2 lb., 2 oz; $5.89
2) 36 oz. = 2 lb., 4 oz; $5.89
3) 53 oz. = 3 lb., 5 oz; $5.89
4) 21 oz. = 1 lb., 5 oz; $3.56

Page 47
1) Small, Size 10
2) Small, Small
3) Large, Large

1) $41.37   2) $63.12
3) $124.54   4) $42.64
5) $58.16   6) 116.67

Page 48, The Key to Simplifying Fractions

Objective: To review the method for simplifying fractions.

Vocabulary: Numerator, denominator, greatest common factor.

ANSWERS:
Top of page:

1) $\frac{1}{2}$ 2) $\frac{4}{5}$ 3) $\frac{7}{8}$ 4) $\frac{3}{5}$

5) $\frac{3}{4}$ 6) $\frac{1}{3}$ 7) $\frac{2}{3}$ 8) $\frac{3}{8}$

9) $\frac{1}{2}$ 10) $\frac{4}{5}$ 11) $\frac{3}{8}$ 12) $\frac{1}{3}$

13) $\frac{1}{3}$ 14) $\frac{3}{4}$ 15) $\frac{2}{3}$ 16) $\frac{2}{9}$

Bottom of page:

1) $1\frac{1}{4}$ 2) $1\frac{4}{7}$ 3) $2\frac{2}{5}$ 4) $1\frac{7}{8}$

5) $1\frac{1}{2}$ 6) $1\frac{5}{8}$ 7) $2\frac{2}{9}$ 8) $1\frac{1}{4}$

9) $1\frac{2}{5}$ 10) $1\frac{4}{5}$ 11) $2\frac{1}{9}$ 12) $2\frac{1}{4}$

13) $4\frac{1}{2}$ 14) $4\frac{1}{4}$ 15) $5\frac{2}{3}$ 16) $5\frac{2}{5}$

Page 49, The Key to Common Denominators

Objective: To find common denominators.
Vocabulary: Factor.

ANSWERS:
1) 6   2) 4   3) 6
4) 6   5) 6   6) 2
7) 2   8) 28   9) 12

Bottom of page:

1) $\frac{5}{8}$, $\frac{4}{8}$ 2) $\frac{4}{8}$, $\frac{3}{8}$ 3) $\frac{5}{15}$, $\frac{3}{15}$

4) $\frac{6}{8}$, $\frac{1}{8}$ 5) $\frac{6}{8}$, $\frac{5}{8}$ 6) $\frac{9}{12}$, $\frac{8}{12}$

7) $\frac{1}{4}$, $\frac{2}{4}$ 8) $\frac{7}{8}$, $\frac{2}{8}$ 9) $\frac{9}{30}$, $\frac{8}{30}$

Page 50, Making Your Own Clothes

Objective: To read a pattern and to compute the amount of fabric required.

Vocabulary: 60″ wide fabric, 44″/45″ wide fabric, 35″/36″ wide fabric, notions, pattern, yard.

Notes: Bring patterns to class. Show various pieces of the pattern. Illustrate how moving the pieces around can save fabric consumption. Review algorithms for adding and subtracting fractions. Discuss the correct operation — adding or subtracting — to use in each exercise.

Page 51
ANSWERS:

1) 2 yards

2) $2\frac{3}{8}$ yards

3) $\frac{1}{8}$ yard

4) $\frac{1}{8}$ yard

5) $1\frac{1}{2}$ yards

6) $3\frac{1}{4}$ yards

7) 2 yards

Page 53

1) $4\frac{3}{4}$ yds.   2) $4\frac{3}{8}$ yds.

3) $4\frac{5}{8}$ yds.   4) $1\frac{3}{8}$ yds.

5) $1\frac{1}{4}$ yds.   6) $\frac{1}{8}$ yd.

7) $\frac{1}{8}$ yd.   8) $\frac{1}{4}$ yd.

9) $6\frac{5}{8}$ yds.   10) $7\frac{1}{2}$ yds.

Page 54, Finding the Cost of Fabric

Objective: To compute the cost of fabric.

Notes: Encourage students to develop their own methods of solution. For example, they could multiply using fractions.

Require students to memorize the decimal equivalents for the fractions in the first exercise. By doing so, they will avoid having to divide each time.

ANSWERS:

1)	.125	2)	.25
3)	.375	4)	.5
5)	.625	6)	.75
7)	.875	8)	1.0
9)	$18	10)	$41.95
11)	$54	12)	$49
13)	$29.51	14)	$159
15)	$28.13	16)	$34.50
17)	$33.71	18)	$70.80
19)	$52.40	20)	$838.81

Page 55, Using a Charge Account

Objective: To compute the new balance of a charge account.

Vocabulary: Charge account, minimum payment, interest, statement.

Notes: Step 2 can be performed in one operation by multiplying by 1.018. Discuss the advantages and disadvantages of using a charge account.

Sample discussion questions:

1. What are some advantages of using a charge account?

2. What are some disadvantages of using a charge account?

3. How can you have the convenience of using a charge card and not have to pay any interest charges?

Page 56

	Unpaid Balance	Interest Charge	New Balance
1)	$235.78	$3.54	$239.32
2)	$86.98	$1.39	$88.37
3)	$68.69	$1.31	$70.00
4)	$2839.76	$56.80	$2896.56
5)	$94.87	$1.71	$96.58
6)	$93.00	$1.21	$94.21
7)	$95.05	$1.62	$96.67
8)	$79.88	$1.20	$81.08
9)	$749.34	$13.11	$762.45
10)	$152.54	$2.29	$154.83

Page 57, Using a Layaway Plan

Objective: To compute the deposit and amount due on a layaway plan.

Vocabulary: Layaway.

ANSWERS:

	Deposit	Due
1)	$1.50	$13.50
2)	$8.00	$16.00
3)	$1.80	$10.20
4)	$10.00	$115.00
5)	$13.20	$52.79

Page 58, Chapter Review

Vocabulary:

A.	Ready-to-Wear	B.	Discounts
C.	Minimum (or down) payment	D.	Layaway

1)	$55.97	2)	$29.01
3)	$9.40	4)	$8.99
5)	$37.57	6)	$.82
7)	$7.78	8)	38%
9)	$2\frac{3}{8}$ yds.		

Chapter 4, MANAGING A HOUSEHOLD (Pages 59-82)

Page 60, Renting a Home

Objective: To apply the Renter's rule.

Vocabulary: Renter's rule, landlord, lease.

Notes: Encourage estimation.
4.3 weeks is an average month since 52 divided by 12 = $4\frac{1}{3}$. For problem #10, give students the hint to find the monthly income first.

ANSWERS:

1)	$279.07	2)	$237.21
3)	$233	4)	$305
5)	$373.02	6)	$192.31
7)	$230.77	8)	$365.38
9)	$280.29	10)	$232.56

Page 61

1) $236
2) $175.60
3) $154.00
4) $192.00
5) $182.40
6) $260

7) $5.00
8) $11.07
9) $14.11
10) $6.97
11) $9.25
12) $8.13

Page 61, Annual Renting Costs

Objective: To compute the cost of rent for one year.

Vocabulary: Annual, 12 months = 1 year.

ANSWERS:

1) $2880
2) $3684
3) $2351.40
4) $3078
5) $5476.32
6) $6942
7) $5866.56
8) $4414.68
9) $3858
10) $2986.80
11) $3300
12) $4560

Page 62, Buying a Home

Objective: To apply the Banker's rule, using annual income.

Vocabulary: Down payment, financed, mortgage, principal, interest.

ANSWERS:

1) $16,475
2) $21,100
3) $26,425
4) $32,175
5) $35,975
6) $39,300
7) $33,900
8) $45,725
9) $51,250
10) $66,875
11) $56,250
12) $42,250

Page 63, Hourly Income

Objective: To apply the Banker's rule, using hourly income.

Note: Forty hours × 52 weeks = 2080 hours worked per year.

ANSWERS:

1) $26,000
2) $41,600
3) $39,000
4) $33,748
5) $27,664
6) $20,800
7) $24,700
8) $26,780
9) $31,980
10) $44,200
11) $49,920
12) $38,220

Note: Estimates may be made using 2000 hours per year. Answers will vary accordingly.

Page 63, Minimum Annual Income

Objective: To use the Banker's rule to calculate the minimum annual income necessary to borrow a given amount of money.

ANSWERS:

1) $14,000
2) $12,000
3) $8,000
4) $16,000
5) $18,000
6) $20,000
7) $22,000
8) $24,000
9) $26,000

Page 64, Computing the Down Payment

Objective: To compute the down payment and amount of a mortgage.

ANSWERS:

	Down Payment	Mortgage
1)	$8,985	$50,915
2)	$5,990	$53,910
3)	$14,975	$44,925
4)	$17,970	$41,930
5)	$9,150	$36,600
6)	$6,628.30	$32,361.70
7)	$19,350	$45,150
8)	$13,131	$59,819
9)	$16,606	$70,794
10)	$11,477	$38,423

Page 65, Computer Application

Objective: To use the computer to solve problems finding down payments and amounts of mortgages.

Notes: Be certain that the program is READY before the students use it. Save it on tape or diskette to minimize preparation time. Rewrite the rates in problems #7 and 8 to 13.5% and 14.5%

ANSWERS:

	Down Payment	Mortgage
1)	$4,574.40	$30,432.60
2)	$11,118.75	$63,006.25
3)	$9,695	$59,555
4)	$7,000.80	$51,339.20
5)	$11,303.50	$75,646.50
6)	$12,144	$63,756
7)	$5,779.35	$37,030.65
8)	$6,887.50	$40,612.50

Page 66, Paying the Mortgage

Objective: To read the monthly payments from the table.

Vocabulary: Fixed-rate mortgage, variable rate, monthly payment.

Notes: Discuss the various mortgage plans. Consult with a banker for the most current information.

ANSWERS:

1)	$506	2)	$713
3)	$771	4)	$664
5)	$717	6)	$796
7)	$747	8)	$1108
9)	Between $630 and $687	10)	Between $829 and $889

Page 67

Objective: To compute the amount repaid on a 30-year mortgage.

Vocabulary: Total interest.

ANSWERS:

	Amount Paid	Total Interest
1)	$217,800	$172,800
2)	$277,560	$202,560
3)	$170,640	$130,640
4)	$239,040	$179,040
5)	$379,760	$249,760
6)	$258,120	$203,120
7)	$249,840	$184,840
8)	$256,680	$206,680
9)	$308,520	$238,520
10)	$273,240	$213,240

Page 68, Terms of Mortgages Differ

Objective: To compute the total amount due on a mortgage.

Page 69

ANSWERS:

1)	$140.00	2)	$407.70
3)	$678.15	4)	$709
5)	$581.94	6)	$576.96
7)	$890.93	8)	$1,543
9)	$1,201.63	10)	$1,155.36

1)	$226,800	2)	$284,472
3)	$345,870	4)	$409,536
5)	$475,146	6)	$541,296
7)	$237,978	8)	$435,780
9)	$333,288	10)	$612,576

Page 70, Utilities

Objective: To read a utility meter.

Vocabulary: Utility, meter, gears, dials.

Notes: Students may be interested in reading their own meters. Have them keep a diary for one week of the daily meter readings in their homes.

Page 71

ANSWERS:

1)	2619	2)	0742
3)	1818	4)	3843
5)	1772	6)	7065
7)	6826	8)	5310
9)	5916	10)	4994

Page 72, Determining Consumption of Utilities

Objective: To compute the number of units and number of cubic feet of a utility consumed.

Vocabulary: Consumed, quarterly, cubic feet, units.

ANSWERS:
1) 17 units, 1700 cubic feet
2) 26 units, 2600 cubic feet
3) 18 units, 1800 cubic feet
4) 36 units, 3600 cubic feet
5) 28 units, 2800 cubic feet
6) 14 units, 1400 cubic feet

Page 73

Objective: To compute the cost of utilities.

Notes: Discuss what happens when the dials reach 9999 + 1. Point out that it occurred between the February and March readings.

Discuss why the consumption varies from month to month and what are some good consumer practices to conserve gas and electricity consumption.

ANSWERS:

	Units	Total Bill
1)	348	$226.20
2)	137	$89.05
3)	652	$423.80
4)	141	$91.65
5)	121	$78.65
6)	135	$87.75
7)	14	$9.10
8)	34	$22.10
9)	29	$18.85
10)	43	$27.95
11)	199	$129.35
12)	119	$77.35

1) 653	$58.77
2) 597	$53.73
3) 562	$50.58
4) 616	$55.44
5) 573	$51.57
6) 516	$46.44
7) 636	$57.24
8) 866	$77.94
9) 1673	$150.57
10) 1370	$123.30
11) 571	$51.39
12) 308	$27.72

Page 74

Objective: To compute average units consumed.

Vocabulary: Average, kilowatt, kilowatt hour, kwh.

Notes: Discuss that computing the average consumption can help a family plan for utility bills.

ANSWERS:

1) 607	2) 456
3) 28	4) 184
5) 644	6) 594

Page 75, Telephone Bills

Objective: To compute telephone expenses.

Notes: Have the class compute the Browns' regular service ($15.10). Discuss telephone service options.

Vocabulary: Flat rate, extension.

ANSWERS:

1) $17.75	2) $18.89
3) $18.98	4) $25.16
5) $16.33	6) $31.74
7) $22.29	8) $21.42
9) $22.16	10) $16.69

Notes: Use the word problems on pages 75 and 76 as a reading exercise. Discuss how to solve story problems.
1) Read the problem carefully.
2) Identify what is to be found.
3) Extract all meaningful numbers.
4) Choose the correct operation.
5) Check that the answer is reasonable.

Page 76

ANSWERS:

1) $1.84	2) $3.60
3) $2.40	4) $5.43
5) $3.34 (Assume the same duration.)	6) $7.95

Page 76

ANSWERS:

1) $110.66	2) $185.33
3) $170.51	4) $133.59
5) $106.02	6) $82.29
7) $97.48	8) $144.87
9) $126.63	10) $82.76
11) $127.63	12) $143.60

Page 77

Objective: To solve story problems concerning utility bills.

Notes: Reinforce story problem solving skills. Discuss each problem with the class. Have students tell how to solve each one. Then have students solve the problems.

ANSWERS:

1) $22.34	2) $80.86
3) $22.34	4) $40.03
5) Electric, $38.69	6) Temperature changes
7) Air conditioning	8) Gas $56.41
9) Elec. $43.29	10) Tel. $26.25

Page 78, Mortgage Insurance

Objective: To compute the amount paid on mortgage insurance claims.

Vocabulary: Mortgage insurance, decreasing benefit term insurance.

Page 79

ANSWERS:

1) $19,600	2) $30,820
3) $8,738	4) $900
5) $6,706	6) $31,680
7) $26,445	8) $32,700
9) $37,440	10) $30,723

Page 80, Homeowners Insurance

Objective: To compute annual payments for homeowners insurance.

Vocabulary: Coverage rate.

Notes: Alert students to the coverage rates. .36% = .0036. Recall the key to changing percents to decimals on page 18.

Page 81

ANSWERS:

1)	$133.20	2)	$286.25
3)	$424.20	4)	$592.50
5)	$324.68	6)	$345.00
7)	$810.00	8)	$393.58
9)	$408.49	10)	$800.70

Page 82, Chapter Review

1) $4158.00
2) $30,000
3) $948.00
4) $341,280.00
5) Use the table on page 68.
$37,900; 15 yrs. at 19% = $114,814.26
$37,900; 40 yrs. at 15% = $227,945.76;
More will be repaid for 40 yrs. at 15%.
6) 6847 units
7) 299 units
8) 310 units
9) $214.13
10) $32,930

Chapter 5, BUYING AND MAINTAINING A CAR (Pages 83-104)

Page 83, New Cars

Objective: To compute the total sticker price.

Vocabulary: Options, base price, transportation/handling.

Notes: Students are usually very eager to share their information about cars. Encourage their participation in this section.

Page 84:

ANSWERS:

1)	$9,064	2)	$10,510

Page 85

3) $8,665
4) $7,744
5) $25,097

Page 86, Used Cars

Objective: To compute the amount saved on car sales.

Vocabulary: Depreciate (the opposite of *appreciate*).

ANSWERS:

1)	$400	2)	$739
3)	$400	4)	$1,770
5)	$510	6)	$880
7)	$540	8)	$2,428

9)	$2,488	10)	$1,602
11)	$2,703	12)	$2,505

Page 87:

Objective: To find the cash price after a trade-in.

Vocabulary: rebate.

ANSWERS: Top of page:

1)	$5,995	2)	$6,549
3)	$4,395	4)	$8,361
5)	$9,452	6)	$5,898
7)	$9,500	8)	$1,999
9)	$9,141	10)	$5,121

Bottom of page:

	With Trade-In	With Rebate
2)	$3,445	$3,695
3)	$4,845	$5,095
4)	$4,645	$4,895
5)	$5,545	$5,795
6)	$3,345	$3,595
7)	$4,145	$4,395
8)	$4,245	$4,495

Page 88, Financing a Car

Objective: To compute the deferred price of a car.

Vocabulary: Deferred price, financing.

ANSWERS:
1)	$4,851	2)	$7,643
3)	$11,971	4)	$15,351
5)	$5,915	6)	$5,239
7)	$7,407		

Page 89

Objective: To compute the difference between the cash price and the deferred price of a car.

Notes: The difference found represents the interest. Discuss the benefits and disadvantages of financing a car.

ANSWERS:
1) $1535.22
2) $1056
3) $1748.24
4) $1542.88
5) $1367.82
6) $2411.30
7) $2234
8) $1647

Page 90, Automobile Insurance

Objective: To compute premiums for auto insurance.

Vocabulary: Liable, liability, premium.

Notes: Be sure that students understand the concept presented with the 10/20/15 notation. The first two numbers refer to the personal injury chart, and the third number refers to property damage chart.

ANSWERS:
1)	$272	2)	$144
3)	$162	4)	$226

Page 91

Objective: To practice solving word problems.

ANSWERS:
1)	$240.72	2)	$414.09
3)	$167.24	4)	$95.94
5)	$732.16		

Page 92, Reading an Odometer

Objective: To write odometer readings in words.

Vocabulary: Odometer, number names.

Notes: Stress the proper use of the word "and." This exercise could be done orally.

ANSWERS:
1) Three hundred four and six tenths miles
2) Seven thousand, one hundred twenty-three and five tenths miles
3) Twelve thousand, eight hundred ninety and five tenths miles
4) Fifty thousand, seventy and one tenth miles
5) Ninety-nine thousand, nine hundred ninety-nine and nine tenths miles
6) Twelve thousand, three hundred forty-five and six tenths miles
7) Thirty thousand, fifty-six and seven tenths miles
8) Seventy-four thousand, five hundred eighty-two, and zero tenths miles
9) Sixty thousand, six and two tenths miles
10) Five thousand, four hundred seventeen and three tenths miles

Page 93, Average Miles Driven per Year

Objective: To compute the average number of miles a car has been driven each year.

Vocabulary: Miles per year. Refer to the Key to Using the Word "Per" on page 34.

ANSWERS:
1)	10,122	2)	8,030
3)	9,951	4)	9,189
5)	2,498	6)	26,813
7)	9,590	8)	9,783
9)	11,701	10)	13,213
11)	13,197	12)	9,882

Page 94, Number of Miles Traveled

Objective: To calculate the number of miles traveled.

ANSWERS:
1)	143.4	2)	91.5
3)	629.9	4)	108.9
5)	3.4	6)	1458.9
7)	930.1	8)	491.3
9)	890.2	10)	897.4

Page 95, Computing Mileage

Objective: To compute miles per gallon or kilometers per liter.

Vocabulary: Mileage, per (see page 34).

Notes: Discuss the factors that influence gas consumption.

ANSWERS:
1) 20 mpg
2) 17 mpg
3) 24 mpg
4) 41 mpg
5) 21 mpg
6) 9 km/L
7) 8 km/L
8) 8 km/L
9) 5 km/L
10) 9 km/L

Page 96, Computing the Range of a Car

Objective: To compute the range of a car in the city and on the highway.

Vocabulary: EPA rating, range.

Notes: This is an approximate range since EPA ratings are a bit inflated.

ANSWERS:

	City	Highway
1)	400 mi.	500 mi.
2)	800 km	1200 km
3)	380 mi.	570 mi.
4)	804 km	1206 km
5)	357 mi.	493 mi.
6)	560 km	880 km
7)	513 mi.	722 mi.
8)	407 km	703 km
9)	522 mi.	666 mi.
10)	351 km	663 km

Page 97, Computing the Fuel Needed

Objective: To compute the amount of fuel needed to travel a given distance.

ANSWERS:
1) 8 gal.
2) 14 gal.
3) 26 gal.
4) 26 gal.
5) 102 gal.
6) 20 L
7) 21 L
8) 47 L
9) 201 L
10) 962 L

Page 98, Computing Average Speed

Objective: To compute the average speed for a given trip.

Vocabulary: mph, km/h

Notes: Discuss factors which influence average speed, such as the number of stops, speed limit, and traffic.

ANSWERS:
1) 40 mph
2) 55 mph
3) 51 mph
4) 20 mph
5) 49 mph
6) 36 km/h
7) 54 km/h
8) 82 km/h
9) 88 km/h
10) 75 km/h

Page 99, Finding Travel Time

Objective: To compute the amount of time necessary for a given trip.

ANSWERS:
1) 4 hours
2) 2 hrs. 36 min.
3) 1 hr. 20 min.
4) 7 hrs. 12 min.
5) 3 hrs. 54 min.
6) 6 hours
7) 13 hrs. 54 min.
8) 4 hrs. 48 min.
9) 24 hrs. 24 min.
10) 30 min.

Page 100, Buying Gasoline

Objective: To compute the cost of gasoline.

Notes: Discuss the various methods of representing the cost per gallon or liter. Discuss the value of and the precision of $1.375.

ANSWERS:
1) $23.38
2) $25.31
3) $26.64
4) $22.08
5) $24.40
6) $28.20
7) $31.65
8) $41.85
9) $34.43
10) $51.15

Page 101

Objective: To determine the savings by using the self-serve gasoline pump.

Vocabulary: Self-serve, full serve.

ANSWERS:
1) $1.60
2) $2.26
3) $1.64
4) $3.88
5) $.96
6) $8.77
7) $.91
8) $4.69
9) $1.00
10) $3.42

Page 102, Repairs

Objective: To compute the total cost for repairs to a car.

Vocabulary: Parts and labor, rebuilt, @.

Page 103:

ANSWERS:

1) $80.95	2) $82.75
3) $125.35	4) $112.25
5) $300.30	6) $25.16
7) $197.18	8) $81.50
9) $160.78	10) $172.61

Page 104, Chapter Review

ANSWERS:

1) $4717.52
2) $239.94
3) Seventy-three thousand, eight hundred four and five tenths miles
4) 202.9 miles
5) $21.84
6) 35 mpg
7) 722 km
8) 30 mph
9) About 7 hours
10) $182.50
11) 12,822
12) 16

Chapter 6, WORKING WITH FOOD (Pages 105-124)

Page 106, Counting Calories

Objective: To calculate the number of calories consumed.

Vocabulary: Calorie, Celsius.

Notes: These calorie counts were obtained from U.S. government publications. Hence the unusual, yet practical problems.

ANSWERS:

1) 430	2) 875
3) 695	4) 761
5) 2761	6) 2000

Page 107:

1) 160 more	2) 425
3) 1635	4) 1488
5) 1575	6) 1995
7) 2720	8) 1720
9) 1650	10) 3755

Page 108, The Key to Ratio

Objective: To express ratios in simplest form.

Vocabulary: Ratio.

Note: Discuss greatest common factor.

ANSWERS:

1) $\frac{1}{2}$ 2) $\frac{80}{1}$

3) $\frac{4}{7}$ 4) $\frac{17}{1}$

5) $\frac{3}{4}$ 6) $\frac{4}{5}$

7) $\frac{1}{2}$ 8) $\frac{40}{1}$

9) $\frac{2}{9}$ 10) $\frac{29}{3}$

11) $\frac{1,000,000}{1}$ 12) $\frac{2}{3}$

13) $\frac{2}{3}$ 14) $\frac{2}{3}$

15) $\frac{2}{3}$ 16) $\frac{2}{3}$

Page 109, The Key to Proportion

Objective: To compute using proportions.

Vocabulary: Proportion, cross product.

Notes: Use cross products by demonstrating that $\frac{3}{4}$ and $\frac{4}{5}$ do not form a proportion.

ANSWERS:

1) ≠	2) ≠	3) ≠
4) =	5) =	6) =
7) ≠	8) ≠	9) ≠

Page 110:

Notes: Encourage students to develop alternate solutions. After all, these are only equivalent fractions.

ANSWERS:

1) 4	2) 8	3) 9
4) 4	5) 80	6) 15
7) 84	8) 30	9) 56
10) 18	11) 8	12) 27
13) 125	14) 16	15) 2

Page 111, Finding Calories with Proportion

Objective: To use proportions to calculate the number of calories in a serving.

Page 112

ANSWERS:

1) 97	2) 450	3) 96
4) 120	5) 180	6) 70
7) 5000	8) 110	9) 408
10) 372	11) 370	12) 475
13) 30	14) 249	15) 499
16) 333	17) 371	18) 186
19) 155	20) 550	21) 2200
22) 440	23) 450	24) 900
25) 600		

Page 113, Recommended Daily Allowances

Objective: To read tables to answer questions about nutrition.

Vocabulary: RDA, nutritionists, nutrient, names of nutrients, I.U., T.

Page 114

ANSWERS:

1) orange juice   2) orange juice
3) carbonated water and ginger ale
4) milk   5) apple juice
6) orange juice
   grape juice
   grapefruit juice, milk
   cranberry and prune juice
   apple juice
   The rest contain no thiamin.
7) prune juice
   apple juice
   grapefruit juice
   cranberry and grape juice
   orange juice
   Whole milk
   The rest contain no iron.

8) cranberry juice
   milk and apple juice
   grapefruit juice
   grape juice
   orange and prune juice
9) grapefruit and orange juice
10) low
11) They contain no vitamins, calcium, or protein.

Page 115

Objective: To use ratios to compute what percent of the RDA of a nutrient is contained in a serving.

ANSWERS:

1) 2%	2) 267%	
3) 8%	4) 24%	
5) 3%	6) 3%	
7) 2%		

Page 115, Computer Application

Notes: To speed up processing, insert these lines:

    45 PRINT
    47 GO TO 10

Remember: Do not key in % signs.

Page 116, Using Calories

Objective: To read the graph to answer questions.

Vocabulary: Basic processes.

ANSWERS:

1) 250	2) 1000	
3) 400	4) 25	
5) 200		

Page 117

Objective: To use proportions to calculate the number of calories consumed in given activities.

Notes: Point out that tables cannot be generated with examples of every case. A general table can be applied to each case using proportions.

In step 1 point out that the ratio used is $\dfrac{\text{minutes}}{\text{calories}}$.

The chart quotes hours of time. In

order to solve the proportion, the time must be in the same units, either hours or minutes. Using hours will introduce fractions. Discuss a problem using fractions. Allow students to choose which method they like.

ANSWERS:

1)	42	2)	75
3)	567	4)	200
5)	56	6)	4625
7)	225	8)	229
9)	229	10)	275

Page 118, Losing Pounds

Objective: To calculate the length of time an activity must be carried out to lose weight.

1)	14 hours	2)	35
3)	14	4)	$4\frac{3}{8}$
5)	56	6)	17.5
7)	7	8)	28

Page 119

Objective: To calculate how many fewer calories per day must be consumed to lose a given number of pounds.

ANSWERS:

1)	500	2)	200
3)	1000	4)	389 (90 days)
5)	750	6)	667
7)	313	8)	233 (90 days)

Page 120, Changing Recipe Yields

Objective: To compute the amounts of ingredients necessary to change recipe yields.

Vocabulary: Yield.

Notes: Call students' attention to the box at the bottom of the page. Take time to refresh or, if necessary, to teach the skill of multiplying fractions. Most students will have learned this operation in an earlier course.

Page 121

Notes: This is a deceptively simple exercise that requires a great deal of organized thought and time. Be alert to possible frustration and plan ahead. Perhaps it would be wise to do one or two examples with the class before asking them to work on their own.

Page 121

ANSWERS:

Number:	1	2	3	4	5	6	7	8	9	10
Ratio:	2	$\frac{1}{2}$	6	$\frac{3}{2}$	$\frac{2}{3}$	3	$\frac{5}{2}$	$\frac{7}{2}$	4	5
Chicken (6)	12	3	36	9	4	18	15	21	24	30
Broccoli (2)	4	1	12	3	$1\frac{1}{3}$	6	5	7	8	10
Soup ($\frac{21}{2}$)	21	$5\frac{1}{4}$	63	$15\frac{3}{4}$	7	$31\frac{1}{2}$	$26\frac{1}{4}$	$36\frac{3}{4}$	42	$52\frac{1}{2}$
Mayonnaise ($\frac{1}{2}$)	1	$\frac{1}{4}$	3	$\frac{3}{4}$	$\frac{1}{3}$	$1\frac{1}{2}$	$1\frac{1}{4}$	$1\frac{3}{4}$	2	$2\frac{1}{2}$
Lemon Juice (1)	2	$\frac{1}{2}$	6	$1\frac{1}{2}$	$\frac{2}{3}$	3	$2\frac{1}{2}$	$3\frac{1}{2}$	4	5
Cheese (4)	8	2	24	6	$2\frac{2}{3}$	12	10	14	16	20

Page 122, Timing Food Preparation

Objective: To calculate the times food must begin cooking to be ready at a given time.

Page 123

ANSWERS:

1) Beef   4:10 P.M.
   Vegetables   7:42 P.M.
   Potatoes   7:20 P.M.
   Rolls   7:43 P.M.
2) Chicken   4:55 P.M.
   Noodles   6:05 P.M.
   Squash   5:30 P.M.
   Biscuits   6:45 P.M.
3) Turkey   9:25 A.M.
   Casserole 3:10 P.M.

   Potatoes   3:00 P.M.
   Beans   3:55 P.M.
4) Stew   3:00 P.M.
   Vegetables 5:55 P.M.
   Dumplings 6:20 P.M.
5) Chicken   5:05 P.M.
   Noodles   4:55 P.M.
   Spinach   4:58 P.M.
   Bread   4:40 P.M.

Page 124, Chapter Review

1) 467 calories   2) $\frac{2}{3}$
3) No   4) 2
5) 96 ounces   6) 267 calories
7) $\frac{3}{4}$ cup   8) 600 calories
9) 167 mL   10) 2:50 P.M.

Chapter 7, IMPROVING YOUR HOME (Pages 125-153)

Page 126, Buying Furniture and Appliances

Objective: To find the original cost of a sale item when the rate of discount is given.

Notes: Allow students to develop their own method for solving these problems. Using proportions is one such alternate method.

Page 127

Notes: Perform one division without rounding and with rounding to show that the error involved is not appreciable.

ANSWERS:

1) $340		11)	$1190
2) $260		12)	$1202
3) $768		13)	$640
4) $230		14)	$750
5) $80		15)	$300
6) $60		16)	$588
7) $370		17)	$200
8) $278		18)	$500
9) $500		19)	$322
10) $341		20)	$260

Page 128

Objective: To compute the rate of discount.

ANSWERS:

1) 40%   2) 38%   3) 29%   4) 33%

Page 129

5) 37%	9) 23%	13) 38%	17) 18%
6) 33%	10) 22%	14) 16%	18) 14%
7) 43%	11) 26%	15) 22%	19) 13%
8) 27%	12) 24%	16) 20%	20) 9%

Page 130, 90-Day Purchase Plan

Objective: To find payment dates on "90 Days Same as Cash" plans.

Vocabulary: 90 Days Same as Cash.

Notes: Review the number of days in each month. Distribute copies of a current calendar for reference.

Page 131

ANSWERS:

1) April 20		11)	February 8
2) June 8		12)	October 28
3) March 30		13)	September 4
4) September 20		14)	March 23
5) August 8		15)	October 8
6) September 26		16)	May 7
7) December 6		17)	March 7
8) November 20		18)	October 2
9) October 14		19)	May 23
10) May 2		20)	August 3

Page 132, The Key to Perimeter and Area

Objective: To apply the rules for computing perimeter and area.

Vocabulary: Perimeter, area.

Notes: Stress the vast differences between perimeter and area. Show examples of figures with the same perimeter but different areas, and vice versa. Discuss the notation, s^2.

Page 133

ANSWERS:

	Perimeter	Area
1)	20 cm	24 cm^2
2)	24"	20 sq. in.
3)	12"	5 sq. in.
4)	20.4 m	20.3 m^2
5)	49.4 cm	150.4 cm^2
6)	80.8 cm	378.9 cm^2
7)	156.6'	1532.5 sq. ft.
8)	12"	9 sq. in.
9)	20 cm	25 cm^2
10)	40 m	100 m^2
11)	25.12 cm	39.44 cm^2
12)	31.76"	63.04 sq. in.
13)	130 m	1056.25 m^2
14)	749.6'	35,118.8 sq. ft.

Page 134, Irregular Shapes

Objective: To compute the area of irregular shapes.

Notes: Students may have difficulty dividing shapes into rectangles. Extra guidance should be given here.

Page 135

ANSWERS:

1) 4 cm^2 2) 30 sq. in.
3) 42 sq. ft.   4) 40 cm^2
5) 44.1 m^2 6) 144 m^2

Page 136, Painting a Room

Objective: To compute the amount of paint needed to cover the walls of a room.

Vocabulary: Coverage.

Notes: For this exercise, disregard the doors and windows. These are approximations.

Page 137

ANSWERS:

1) 2 quarts
2) 3 gallons
3) A diagram should be drawn.

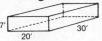

4) 140 sq. ft., 210 sq. ft., 140 sq. ft., 210 sq. ft.
5) 700 square feet
6) 10 gallons

---

1) 3 quarts
2) 4 quarts
3) 5 quarts
4) 7 quarts
5) 7 quarts

Page 138, Buying Paint

Objective: To compute the cost of paint.

Vocabulary: 4 quarts = 1 gallon.

Page 139

ANSWERS:

	Quarts Required	Amount to Buy		Cost		
		Gallons	Quarts	Gallons	Quarts	Total
1)	9	2	1	$23.98	$3.99	$27.97
2)	10	2	2	$23.98	$7.98	$31.96
3)	6	1	2	$11.99	$7.98	$19.97
4)	7	1	3	$11.99	$11.97	$23.96
5)	13	3	1	$35.97	$3.99	$39.96
6)	46	11	2	$131.89	$7.98	$139.87

Page 140, Buying Wallpaper

Objective: To compute the cost of wallpaper.

Vocabulary: Double roll, single roll, wall area.

Notes: Cut out a mock-up of the room from tagboard. Open the four walls to illustrate the rule.

Page 141

ANSWERS:

1) 2   2) 3
3) 3   4) 4
5) 4   6) 3
7) 3   8) 4
9) 3   10) 5

Page 142, Covering the Floor

Objective: To compute the cost of tiling a floor.

Notes: Some tiles are sold in 10" or 9" squares.

ANSWERS:
1) $62.30   2) $158.40
3) $113.52   4) $616.59
5) $388.50   6) $1,343.34
7) $332.64   8) $549.27
9) $386.64   10) $483.36

Page 143

Objective: To compute the cost of carpeting.

Page 144

ANSWERS:
1) $119.85   6) $125.94
2) $104.83   7) $32.34
3) $265.90   8) $79.95
4) $32.55   9) $89.95
5) $49.17   10) $295.35

Page 144, Computing Length of Molding

Objective: To find the amount of molding required for a given room.

ANSWERS:
1) 34'   2) 52'
3) 38'   4) 60'
5) 50'   6) 70'
7) 54'   8) 52'
9) 60'   10) 70'

Page 145

1) 40' 2"   6) 59' 4"
2) 66' 8"   7) 37' 4"
3) 81' 10"   8) 57' 10"
4) 37' 10"   9) 56' 8"
5) 41'   10) 88' 6"

Page 146, Wall-to-Wall Carpeting

Objective: To estimate the cost of carpet for a given room.

Notes: Answers will vary depending on how students proceed with their solutions.

ANSWERS:
1) $54   2) $160
3) $550   4) $171
5) $250   6) $104
7) $220   8) $351
9) $480   10) $336

Page 147, Adding On

Objective: To compute the cost of additions to a house.

Vocabulary: Contractors, materials.

Notes: Discuss the factors which influence the cost per square foot.

ANSWERS:
1) $8,880   6) $1,825
2) $3,650   7) $3,637.50
3) $21,200   8) $2,953
4) $2,151   9) $4,075.40
5) $9,609   10) $12,575

Page 148, Insulation

Objective: To compute the cost of insulating an attic.

ANSWERS:
1) $.20   2) $.20
3) $.29   4) $.33

Page 149:

1) $122.43   2) $187.53
3) $419.70   4) $236.85
5) $262.35   6) $195.86

An extra problem for students:

For some of these homes there was a great deal of waste. Too much insulation was purchased. Keep in mind that you have to purchase a complete roll of insulation. Mix brands if you like, but see if you can insulate these six homes for less money than was spent in exercises 1 to 6.

Page 150, Seeding and Feeding Lawns

Objective: To find the cost of upkeep for lawns.

Vocabulary: Established lawns, thin lawns, m².

ANSWERS:
1) $.04   2) $.04
3) $.02   4) $.02
5) $.01

Page 151

	Seed	Fertilizer
1)	$47.16	$80.82
2)	$37.95	$80.82
3)	$19.98	$62.86
4)	$42.45	$179.60
5)	$22.76	$134.70

Page 152, Fencing In a Yard

Objective: To find the length of fencing required.

Notes: Ask students why this is a perimeter problem rather than an area problem.

ANSWERS:

1)	72 m	2)	80 m
3)	132 m	4)	58.8 m
5)	90.2 m	6)	259.4 m
7)	139.2 m	8)	133.4 m
9)	113.2 m	10)	148.2 m

Page 153, Chapter Review

ANSWERS:

1)	$100	2)	43%
3)	January 8	4)	2 cans
5)	9 rolls	6)	20 sq. yds.
7)	527.01	8)	900
9)	534 feet		

Chapter 8, TRAVELING (Pages 154-186)

Page 155, Reading a Map

Objective: To read information from a map.

Vocabulary: Legend.

ANSWERS:

1)	46	4)	2
2)	8	5)	12 miles
3)	Yes	6)	No

Page 157

ANSWERS:

1)	7	10)	18
2)	16	11)	18
3)	33	12)	About 23 miles Rtes. 947 and 9
4)	26 or 22	13)	26
5)	13	14)	73N, 195W, 34W, 1N, 24 miles
6)	17		
7)	15	15)	947W, 19W, 222S, 20 miles
8)	7		
9)	5		

Page 159, Estimating Distances

Objective: To use the scale on a map to estimate distances.

Vocabulary: Scale, straightline distance.

Notes: Students will need rulers for this activity. They measure to the nearer $\frac{1}{4}$". If the students are un-comfortable with rapid changes from fractions to decimals to fractions, the entire activity can be done using fractions.

According to the map on page 156, the distance from Baltiless to Fallfield is 30 miles. The estimated distance is greater because distances on page 156 are rounded to the nearer mile. Both of these methods yield only approximate distances. The road distance might be greater in some cases since the road does not follow a straight line.

ANSWERS:

	Distance In Inches	Distance In Miles
1)	$1\frac{1}{2}$ "	9 miles
2)	$6\frac{3}{4}$ "	$40\frac{1}{2}$ miles
3)	$5\frac{1}{2}$ "	33 miles
4)	$5\frac{1}{4}$ "	$31\frac{1}{2}$ miles
5)	$4\frac{3}{4}$ "	$28\frac{1}{2}$ miles
6)	$2\frac{1}{4}$ "	$13\frac{1}{2}$ miles
7)	3"	18 miles

Page 161, Using a Mileage Diagram

Objective: To compute distances using a mileage diagram.

Vocabulary: Mileage diagram.

Notes: Students will find it easier to copy the chart on grid paper. Allow the students who notice that the top half of the grid is a mirror image of the bottom half to complete only the bottom half. Discuss the shortcomings of this map, and its advantages.

ANSWERS:

1)	941
2)	985	1086
3)	640	818	345
4)	2003	1658	1018	1363
5)	1935	994	1092	1370	1028
6)	3165	2548	2180	2525	1162	1554
7)	1584	643	1502	1461	2074	1236	2790
8)	1134	500	586	532	1158	894	2320	916
9)	1577	636	990	1068	1386	358	1912	878	536
10)	211	730	831	486	1849	1724	3011	1373	923	1366
11)	3270	2925	2285	2630	1267	1957	403	3193	2425	2315	3116
12)	3016	2897	2031	2376	1377	2405	1261	3313	2397	2763	2862	858
13)	1209	795	291	569	863	801	2025	1211	295	699	1030	2130	2102
14)	437	504	687	396	1667	1498	2829	1147	697	1140	226	2934	2718	804

Page 162, Reading a Train Schedule

Objective: To use a train schedule to compute the length of time for a trip.

Vocabulary: READ UP, READ DOWN

Notes: Bring sample train schedules to class. Show the students how to read them. Remind the students about the 24-hour clock.

Page 163

ANSWERS:

1) 7 hrs., 6 min.   2) 6 hrs., 37 min.
3) 1 hr., 8 min.   4) 2 hrs., 55 min.
5) 2 hrs., 48 min.   6) 1 hr., 11 min.
7) 5 hrs., 14 min.   8) 1 hr., 58 min.
9) 5 hrs., 53 min. 10) 32 min.

11) 10 min.
12) 10 min.
13) 15 min.
14) 20 min.

Page 164, Computing Train Fares

Objective: To compute the train fare for a given trip.

Vocabulary: Coach, parlor, excursion.

Page 165

ANSWERS:

1)	$26.50	6)	$109.13
2)	$18.00	7)	$100.75
3)	$59.50	8)	$16.00
4)	$95.00	9)	$109.73
5)	$5.25	10)	$20.00

Page 166, Staying in Hotels

Objectives: To calculate the cost of renting a room in a hotel.

Vocabulary: Single, double, suite, tourist season.

Page 167

ANSWERS:

	Number Of Days	Daily Rate	Total Cost	Cost Per Person
1)	7	$29.50	$206.50	$206.50
2)	1	24.50	24.50	12.25
3)	8	43.50	348.00	174.00
4)	10	67.50	675.00	135.00
5)	9	57.50	517.50	129.38
6)	6	43.50	261.00	130.50
7)	4	23.50	94.00	94.00
8)	11	29.50	324.50	162.25
9)	14	67.50	945.00	315.00
10)	6	85.00	510.00	170.00

Extra Problems:

Some hotel guests leave a tip for the maid who straightens the room. Compute 15% of each of the costs per person in exercises 1 through 10.

ANSWERS:
1) $30.98
2) $1.84
3) $26.10
4) $20.25
5) $19.41
6) $19.58
7) $14.10
8) $24.34
9) $47.25
10) $25.50

Page 168

Notes: Discuss the need to add 1 day. "Subtracting is 'taking away,' and we do not want to take away the 29th day." Use the calendar on page 31 if you like.

ANSWERS:

	Days	Daily Rate	Total Cost	Cost per Person
1)	10	$29.50	$295	$147.50
2)	6	85	510	102
3)	10	24.50	245	122.50
4)	20	19.50	390	390
5)	12	95	1140	162.86

Page 169

Notes: Students will need three lines to record the answers to each problem in the chart.

ANSWERS:

	Season	No. of Days	Daily Rate	Total Cost	Cost per Person
1)	1	5	$95	$475	
	2	5	67.50	337.50	
				812.50	$270.83
2)	1	8	29.50	236	
	2	5	24.50	122.50	
				358.50	179.25
3)	1	5	29.50	147.50	
	2	3	39.50	118.50	
				266.00	266.00
4)	1	12	29.50	354.00	
	2	3	24.50	73.50	
				427.50	213.75
5)	1	5	57.50	287.50	
	2	26	85.00	2210.00	
				2497.50	499.50

Page 170, Package Travel Plans

Objective: To compute the cost per day of package plans.

Vocabulary: Package plans.

Notes: Bring travel brochures to class.

ANSWERS:
1) $113.57
2) $23.33
3) $99.86
4) $86.13
5) $149.14
6) $142.00
7) $187.38
8) $169.75
9) $217.00
10) $156.88

Page 171, Exchanging Currency

Objective: To find the value of American dollars in other currencies.

Note: Exchange rates vary daily. These are sample rates. Consult a recent conversion table for more up-to-date exchange rates.

Page 172

ANSWERS: Middle of page:
1) 64 dollars
2) 1,512 schillings
3) 2800 francs
4) 5,670 cruzeiros
5) 42 pounds
6) 85 dollars
7) 106 yuan
8) 539 krones
9) 314 markkas
10) 510 francs
11) 4200 drachmas
12) 410 rupees
13) 621 shekels
14) 103,670 lire
15) 18,900 yen
16) 1876 pesos
17) 401 kronor
18) 155 francs
19) 181 marks

ANSWERS: Bottom of page:
1) 60 pounds
2) 61 dollars
3) 670 pesos
4) 103.2 marks
5) 125,885 lire
6) 269.73 francs
7) 638.73 yuan
8) 93,135 shekels
9) 215,460 yen
10) 1,500,000 drachmas

Page 173

Objective: To find the value of a currency in American dollars.

ANSWERS:
1) $28.34		10) $6.18
2) $18.24		11) $221.00
3) $168.85		12) $106.37
4) $309.85		13) $81.53
5) $124.50		14) $27.40
6) $41.00		15) $6.50
7) $265.98		16) $117.00
8) $128.31		17) $1,222.20
9) $22.00		

Page 174, Renting a Car

Objective: To compute the cost of renting a car.

Vocabulary: Sub-compact, compact, mid-size, full size.

Notes: Referring to page 175, in Weekly Plan I, the daily rate is $19.95 plus $3.50 for insurance. This is unclear from the rate chart and should be pointed out to the students.

Choose a month and an amount of miles. Compute the various rental fees for that situation. Discuss which rental plan is best and under what conditions another plan might be better.

Page 176

1) $180.55		6) $2885.80
2) $458.00		7) $202.00
3) $492.45		8) $3931.80
4) $129.00		9) $656.60
5) $876.00		10) $764.00

	Hurts	Bevis	Carl Eece 1	2
11)	$189	$432.86	$164.15	$236
12)	324	289.50	Daily:	216.60
13)	458	569.30	328.30	310
14)	150	166.45	Daily:	125.15
15)	1832	1814.60	1313.20	2180.14

Page 177, Parking Expenses

Objective: To apply the parking rate chart to a given time.

ANSWERS:
1) $1.75		2) $2.50
3) $3.25		4) $4.00
5) $4.75		6) $5.00

7) $4.75		8) $3.25
9) $4.00		10) $2.50

Page 178

Objective: To compute elapsed time and parking costs.

Notes: Discuss the 24-hour clock with the students. Try a parking example that spans the midnight hour. Discuss how Max may charge customers, (i.e., as a continuation of the same day or as a new day).

Page 179

ANSWERS:
1) $4.25		2) $3.25
3) $3.25		4) $3.25
5) $4.25		6) $2.25
7) $4.25		8) $6.00
9) $6.00		10) $6.00

Page 180, Time Zones

Objective: To find the time in one time zone when the time is known in another zone.

Page 182

ANSWERS:
1) 3:00 P.M.		6) 11:48 A.M.
2) 8:25 A.M.		7) 2:15 A.M.
3) 12:04 P.M.		8) 12:36 A.M.
4) 3:59 A.M.		9) 12:01 P.M.
5) 12:37 A.M.		10) 6:52 P.M.

Page 183, Traveling By Air

Objective: To calculate the length of time for a flight.

ANSWERS:
1) 3 hrs., 11 min		2) 2 hrs., 45 min.
3) 2 hrs., 5 min.		4) 3 hrs., 15 min.
5) 4 hrs., 15 min.		6) 4 hrs., 37 min.
7) 4 hrs., 30 min.		8) 7 hrs., 45 min.
9) 9 hrs., 10 min.		10) 4 hrs., 33 min.

Page 184

Objective: To calculate the length of time for a flight which spans two or more time zones.

Note: The dash in the chart indicates that the city is in that time zone.

Page 185

Notes: This is a very challenging exercise. Be *sure* students are capable of proceeding on their own before allowing them to work alone. Discuss factors that affect times on out-going trips versus return trips.

ANSWERS:

1) 9 hrs., 55 min.   5) 2 hrs., 44 min.
2) 1 hr., 52 min.   6) 6 hrs., 15 min.
3) 4 hrs., 53 min.   7) 4 hrs., 23 min.
4) 6 hrs., 27 min.   8) 14 hrs., 10 min.

9) 3 hrs., 50 min.   18) 1 hr., 56 min.
10) 5 hrs., 40 min.   19) 1 hr., 35 min.
11) 4 hrs., 45 min.   20) 7 hrs.
12) 3 hrs., 45 min.   21) 7 hrs., 55 min.
13) 3 hrs., 25 min.   22) 9 hrs., 20 min.
14) 1 hr., 48 min.   23) 9 hrs., 45 min.
15) 2 hrs., 5 min.   24) 13 hrs., 5 min.
16) 3 hrs., 55 min.   25) 4 hrs., 30 min.
17) 4 hrs., 5 min.

Page 186, Chapter Review

1) 26   6) $108
2) 30 km; 12.5 km   7) $174.59
3) 2 hrs., 19 min.   8) $115.39
4) $36.56   9) $5.50
5) 14 nights   10) $5:45 P.M.

Chapter 9, BUDGETING YOUR MONEY (Pages 187-200)

Page 188, Finding Average Income

Objective: To compute average income.

ANSWERS:

1) $775.32   2) $564.94
3) $294.58   4) $690.39
5) $386.58   6) $622.18
7) $557.92   8) $676.22
9) $753.78   10) $853.15

Page 189

1) $1030.44   2) $512.65
3) $473.00   4) $1075.48
5) $572.00   6) $335.10
7) $613.87   8) $411.08
9) $1370.55   10) $302.14

Page 190, Preparing a Budget

Objective: To determine which expenses are included in a budget.

ANSWERS: Answers will vary.

1) Rent or mortgage payment, repairs, insurance on home, utility bills.
2) Meals at home, school lunches, eating out.
3) New clothing, cleaning costs, alterations.
4) Purchase of a car, gas, oil, parking, repairs, bus fares, taxi fares.
5) Life insurance, health insurance, auto insurance.
6) Birthdays, holidays, showers, weddings.
7) Movies, theater, sports events.
8) Savings accounts, money market funds, investments.
9) Doctor and dentist bills, prescriptions.
10) Newspapers, magazines, haircuts, charitable contributions.

Page 191

Objective: To compute the amount of money to be budgeted for each category.

ANSWERS:

1)	2)	3)
$218.74	$122.87	$134.04
$162.03	$91.01	$99.29
$72.91	$40.96	$44.68
$105.32	$59.16	$64.54
$40.51	$22.75	$24.82
$56.71	$31.85	$34.75
$40.51	$22.75	$24.82
$64.81	$36.41	$39.72
$24.30	$13.65	$14.89
$24.30	$13.65	$14.89

4)	$128.57	5)	$268.06	6)	$231.43
	$95.24		$198.56		$171.43
	$42.86		$89.35		$77.14
	$61.91		$129.06		$111.43
	$23.81		$49.64		$42.86
	$33.33		$69.50		$60.00
	$23.81		$49.64		$42.86
	$38.10		$79.42		$68.57
	$14.29		$29.78		$25.71
	$14.29		$29.78		$25.71

7)	$192.59	8)	$91.05	9)	$304.57
	$142.66		$67.44		$225.61
	$64.20		$30.35		$101.52
	$92.73		$43.84		$146.64
	$35.66		$16.86		$56.40
	$49.93		$23.60		$78.96
	$35.66		$16.86		$56.40
	$57.06		$26.98		$90.24
	$21.40		$10.12		$33.84
	$21.40		$10.12		$33.84

10)	$136.37	11)	$280.18	12)	$168.07
	$101.02		$207.54		$124.50
	$45.46		$93.39		$56.02
	$65.66		$134.90		$80.92
	$25.25		$51.88		$31.12
	$35.36		$72.64		$43.57
	$25.25		$51.88		$31.12
	$40.41		$83.02		$49.80
	$15.15		$31.13		$18.67
	$15.15		$31.13		$18.67

Page 192, Adjusting a Budget

Objective: To compute the percent of income spent on each item of a budget.

Page 193

ANSWERS:

1) a) 32%   2) 26%
   b) 24%   21%
   c) 10%   9%
   d) 15%   14%
   e) 5%   6%
   f) 4%   8%
   g) 3%   5%
   h) 1%   7%
   i) 6%   4%

3) Records, 8%
   Savings, 66%
   Entertainment, 18%
   Hair, 8%

Page 194

4) a) 11%
   b) 3%
   c) 15%
   d) 54%
   e) 4%
   Profit, 13%

Page 194, Using Circle Graphs

Objective: To use circle graphs to answer questions about budgets.

Notes: Students will need protractor use skills before doing this section.

Page 195

ANSWERS:

1) a) 90°
   b) 45°
   c) 27°
   d) 32°
   e) 61°
   f) 23°

2) Answers will vary.
   Large family, more than one car.

3) Food   86°
   Housing   54°
   Clothing   72°
   Cars   36°
   Health   36°
   Miscellaneous   76°

4) Housing   25%   90°
   Food   20%   72°
   Clothes   15%   54°
   Transportation 13%   47°
   Investments   12%   43°
   Health   5%   18°
   Miscellaneous   10%   36°

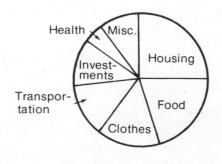

ANSWERS:

	Food	Rent	Clothes	Cars
June Balance	$ 4.10	0	13.50	3.15
+ July Budget	+ 175.65	310.50	25.00	70.00
Funds Available	_179.75_	_310.50_	_38.50_	_73.15_
– July Expenses	– 176.50	310.50	18.00	67.36
July Balance	_3.25_	_0_	_20.50_	_5.79_
+ August Budget	_+175.65_	_310.50_	_25.00_	_70.00_
Funds Available	_178.90_	_310.50_	_45.50_	_75.79_

	Gifts	Savings	Miscellaneous
June Balance	$36.50	0	176.10
+ July Budget	+ 20.00	45.00	57.50
Funds Available	_56.50_	_45.00_	_233.60_
– July Expenses	– 10.99	45.00	65.60
July Balance	_45.51_	_0_	_168.00_
+ August Budget	_+20.00_	_45.00_	_57.50_
Funds Available	_65.51_	_45.00_	_225.50_

Page 197
ANSWERS:
1) $703.65   2) $693.95   3) 99%

	Food	Shelter	Clothes
	25%	30%	20%
January Balance	$ 1.15	11.65	0
+ February Budget	_+243.84_	_292.61_	_195.07_
Funds Available	_244.99_	_304.26_	_195.07_
– February Expenses	– 236.36	303.50	86.58
February Balance	_8.63_	_.76_	_108.49_

	Cars	Savings	Miscellaneous
	18%	5%	2%
January Balance	36.09	0	82.74
+ February Budget	_+175.56_	_48.77_	_19.51_
Funds Available	_211.65_	_48.77_	_102.25_
– February Expenses	– 111.06	48.77	97.36
February Balance	_100.59_	_0_	_4.89_

To find the total amount spent in January, add the amounts of the January balances; then subtract the sum from $975.35. The total is $843.72.

Page 198
Objective: To compute funds available when a budget is overspent.

Note: Students may be unfamiliar with negative numbers. Discuss the notation of the negative sign.

Page 199
ANSWERS:
1) He could borrow from his savings account. Or, perhaps he had an extra $6.90 in one of his other budget categories. This is popularly known as "robbing Peter to pay Paul."
2) – $13.85   5) $8.16 in debt (each)
3) $3.75   6) $26.42
4) $36.43   7) $93.70

Page 200, Chapter Review
1) $1069.80
2) $86.66
3) 30%
4) Housing 97°
   Food 86°
   Cars 68°
   Clothes 54°
   Health 29°
   Savings 18°
   Miscellaneous 7°

The total is 359° instead of 360° due to rounding.

5)

	Housing	Food	Clothes
	28%	25%	10%
Old Balance	$16.76	–$10.50	–$3.94
+ New Budget	_+196.00_	_175.00_	_70.00_
Funds Available	_212.76_	_164.50_	_66.06_
– Expenses	– 200.00	186.40	57.87
New Balance	_12.76_	_–21.90_	_8.19_

	Cars	Health	Savings and Miscellaneous
	15%	2%	20%
Old Balance	$111.50	$36.88	$47.89
+ New Budget	_+105.00_	_14.00_	_140.00_
Funds Available	_216.50_	_50.88_	_187.89_
– Expenses	– 235.84	10.94	100.00
New Balance	_–19.34_	_39.94_	_87.89_

An extra activity for the students:
Prepare a budget for yourself. Account for all your expenses and express the budget allotments in terms of a percent of your net income.

Chapter 10, BANKING AND INVESTING (Pages 201-226)

Page 202, Simple Interest

Objective: To compute simple interest for time in years.

Notes: Simple interest is rarely used. However, it is a good introduction to compound interest, which is used extensively.

ANSWERS:

	Interest	Total
1)	$30	$330
2)	$27	$177
3)	$65.80	$723.80
4)	$4.40	$26.40
5)	$83.20	$211.20
6)	$262.50	$1262.50
7)	$1.55	$11.55
8)	$12.50	$110.50
9)	$547.81	$2074.81
10)	$34.27	$77.27

Page 203

Objective: To compute simple interest for time in months.

Notes: To avoid confusion in Step 1, one could change the fraction $\frac{3}{8}$ to a decimal first: 5.375%. Then change the percent to a decimal: .05375.

ANSWERS:

	Interest	Total
1)	$2.75	$102.75
2)	$22.97	$522.97
3)	$11.25	$611.25
4)	$42.75	$942.75
5)	$191.25	$1691.25
6)	$645.31	$7645.31
7)	$1209.45	$10,005.45
8)	$622.31	$4645.31
9)	$1093.75	$11,093.75
10)	$5,437.50	$25,437.50

Page 204, Compound Interest

Objective: To compute compound interest and the new balance using the formula, $I = P \times R \times T$.

Notes: After practicing the 4-step method on page 204, suggest that steps 1 and 2 and steps 3 and 4 may be combined by multiplying by 1.025.

Some banking institutions declare that the year is 360 days long to simplify computations. Whether students use 360 or 365 will not cause great variation in the result. Due to possible confusion, no problems of that type were included.

Page 205

ANSWERS: Answers will vary a few cents, depending on how often rounding is performed.

1)	$106.09	8)	$316.84
2)	$107.12	9)	$422.46
3)	$105.58	10)	$528.07
4)	$106.35	11)	$1,020.06
5)	$107.64	12)	$5,096.14
6)	$105.61	13)	$1,540.24
7)	$211.24		

Computer Application

On page 35 of this Guide is a program that will calculate compound interest and the new principal.

Page 206

Objective: To compute the value of an investment subject to compound interest using a table.

ANSWERS:

	Compounded Annually	With Simple Interest
1)	$12,017.50	$6,750.00
2)	$57,765.00	$11,000.00
3)	$126,645.00	$13,125.00
4)	$230,433.00	$33,250.00
5)	$438,683.00	$38,500.00
6)	$23,242.05	$15,892.00
7)	$19,277.20	$10,657.70
8)	$607,520.00	$93,600.00
9)	324,136.80	$145,080.00
10)	$658,091.04	$203,352.00

COMPUTER APPLICATION

```
10   REM ****** COMPOUND INTEREST ******
20   PRINT "THIS PROGRAM WILL CALCULATE COMPOUND
     INTEREST."
30   PRINT
40   PRINT "WHAT IS THE PRINCIPAL? ";
50   INPUT P
60   PRINT "WHAT IS THE RATE OF INTEREST? ";
70   INPUT R
80   R1 = R
90   PRINT "HOW MANY YEARS WILL THE INTEREST BE
     COMPOUNDED? "
100  INPUT N
110  N1 = N
120  PRINT "SELECT THE NUMBER OF TIMES THE INTEREST
     IS COMPOUNDED:"
130  PRINT TAB(10) "(1)   ANNUALLY"
140  PRINT TAB(10) "(2)   SEMIANNUALLY"
150  PRINT TAB(10) "(3)   QUARTERLY"
160  PRINT TAB(10) "(4)   MONTHLY"
170  PRINT TAB(10) "(5)   DAILY"
180  INPUT T
190  IF T = 2 THEN N = N * 2 : R = R / 2
200  IF T = 3 THEN N = N * 4 : R = R / 4
210  IF T = 4 THEN N = N * 12 : R = R / 12
220  IF T = 5 THEN N = N * 365.25 : R = R / 365.25
230  IF T < 1 OR T > 5 PRINT "SELECT A NUMBER BETWEEN 1
     AND 5.":  GOTO 120
240  I = P * ( 1 + R/100) ↑ N
250  I = INT (I * 100 + .5) /100
260  S$(1)="ANNUALLY":S$(2)="SEMIANNUALLY":
     S$(3)="QUARTERLY":S$(4)="MONTHLY":S$(5)="DAILY"
270  PRINT "THE INTEREST COMPOUNDED ";S$(T);" ON
     $";P;" AT ";R1;"% FOR ";N1;
280  PRINT " YEARS IS ";
290  PRINT TAB(20) "$";I−P
300  PRINT
310  PRINT "THE NEW PRINCIPAL IS $";
320  PRINT I
330  END
```

Page 208, Doubling Your Money

Objective: To apply the Rule of 100 and the Rule of 72.

Page 209

ANSWERS:

1)	20	2)	10	3)	4
4)	25	5)	50	6)	100
7)	2	8)	2.5	9)	$6\frac{2}{3}$
10)	$12\frac{1}{2}$	11)	$8\frac{1}{3}$	12)	$9\frac{1}{11}$
13)	12	14)	6	15)	9
16)	8	17)	18	18)	24
19)	4	20)	3	21)	7.2
22)	14.4	23)	3.6	24)	$6\frac{6}{11}$

Page 211

Objective: To write amounts of checks in words.

Notes: To prepare students to write the amounts, have them take turns reading the amounts at the bottom of page 211 aloud. Be alert to a misuse of the word "and."

ANSWERS:

1) Twenty-five and $\frac{86}{100}$

2) Thirty-seven and $\frac{16}{100}$

3) One hundred forty-three and $\frac{No}{100}$

4) Nine hundred five and $\frac{15}{100}$

5) One thousand three hundred twenty-seven and $\frac{56}{100}$

6) Forty-eight and $\frac{18}{100}$

7) Two hundred ninety-six and $\frac{75}{100}$

8) Four hundred forty-nine and $\frac{37}{100}$

9) Fifty-seven and $\frac{49}{100}$

10) Eighteen and $\frac{18}{100}$

11) Sixty-one and $\frac{60}{100}$

12) Seventy and $\frac{61}{100}$

13) Forty-five and $\frac{76}{100}$

14) Eighty-two and $\frac{72}{100}$

15) Three hundred eighty-four and $\frac{48}{100}$

Page 212, Keeping the Account Up-to-Date

Objective: To compute the new balance of a checking account.

Vocabulary: Transaction, register, stub.

Page 213

Notes: An error on any exercise will cause all succeeding answers to be incorrect. Devise a self-checking mechanism so that students will keep going with correct results.

ANSWERS:

1)	$452.25	11)	$102.83
2)	$368.13	12)	$257.93
3)	$330.68	13)	$161.15
4)	$320.68	14)	$130.05
5)	$475.78	15)	$117.49
6)	$378.39	16)	$127.49
7)	$227.60	17)	$90.99
8)	$382.70	18)	$246.09
9)	$353.72	19)	$240.61
10)	$313.72	20)	$220.61

Page 214, Reconciling a Checking Account

Objective: To reconcile a checking account.

Vocabulary: Reconcile, processed, unprocessed, monthly statements.

Lab Activity: Provide students with a check register, returned checks, and a bank statement. Have them reconcile the account. This can be done with as few as five items in the register and one or two returned checks.

Page 215

ANSWERS:
1) 10¢
2) 50¢
3) $4.46

Page 216
4) Balances
5) $1.99
6) Balances
7) 8¢

8) $16.56
9) $10.10
10) $.28

Page 217, Stock Market Mathematics
Objective: To compute the price of a stock after an increase.

Page 218
Notes: Show students the stock market reports in the newspaper. Discuss the prices of stocks for well-known or local companies. Discuss the symbols used for stock names in the listings.

ANSWERS:

1) $96\frac{1}{8}$ 7) $22\frac{5}{8}$

2) $28\frac{3}{8}$ 8) $20\frac{3}{8}$

3) $8\frac{1}{8}$ 9) $94\frac{3}{4}$

4) $11\frac{5}{8}$ 10) $26\frac{1}{2}$

5) 19   11) $27\frac{1}{4}$

6) $40\frac{3}{4}$

Page 219
Objective: To compute the profit from stock purchases.

ANSWERS:

1) $5\frac{5}{8}$ 5) $4\frac{1}{2}$

2) $\frac{3}{4}$ 6) $3\frac{7}{8}$

3) $1\frac{1}{4}$ 7) $3\frac{5}{8}$

4) $10\frac{7}{8}$ 8) $4\frac{1}{2}$

Page 220
Objective: To compute the profit or loss from stock purchases.

1) $5\frac{5}{8}$ 6) $3\frac{7}{8}$

2) $1\frac{1}{4}$ 7) $4\frac{1}{8}$

3) $1\frac{1}{4}$ 8) $3\frac{7}{8}$

4) $2\frac{1}{2}$ 9) $4\frac{1}{2}$

5) $2\frac{1}{8}$ 10) $3\frac{5}{8}$

11) $-2\frac{1}{4}$ 16) $-1\frac{1}{4}$

12) -1 17) $-38\frac{3}{8}$

13) $-\frac{1}{2}$ 18) $-\frac{1}{2}$

14) $+1\frac{1}{4}$ 19) $-10\frac{7}{8}$

15) $+\frac{5}{8}$ 20) $+5\frac{3}{8}$

Page 221
Objective: To compute the cost of round lots of shares of stock.

Vocabulary: Round lots.

Note: The chart on page 217 could be used to convert eighths of a dollar into cents.

ANSWERS:

1) $1,237.50   2) $531.25
3) $10,450.00   4) $1,350.00
5) $4,275.00   6) $14,062.50
7) $1,687.50   8) $4,275.00
9) $2,325.00   10) $12,400.00
11) $7,450.00   12) $45,243.75
13) $39,812.50   14) $1,750.00
15) 132,962.50   16) $43,987.50
17) $25,800.00   18) $18,000.00
19) $14,750.00   20) $23,250.00

Page 222
Objective: To compute the number of shares that can be purchased for a given amount of money.

Note: Some divisions could be done using fractions instead of decimals.

ANSWERS:

	Number	Cost
1)	76	$494
2)	153	$3997.13
3)	49	$992.25
4)	23	$97.75
5)	117	$1491.75
6)	1133	$499.88
7)	130	$1998.75
8)	120	$75
9)	82	$799.50
10)	58	$1189
11)	369	$599.63
12)	64	$296
13)	238	$2499
14)	200	$900

15)	923	$2999.75
16)	114	$9975
17)	107	$1792.25
18)	319	$6499.63
19)	223	$5993.13
20)	465	$7498.13

Page 223, Evaluating Profits and Losses

Objective: To compute the percent of increase or decrease of stock price.

Vocabulary: Return on money, "realize" a profit or loss.

Page 224

ANSWERS:

1)	Loss, 31%	2)	Profit, 63%
3)	Loss, 31%	4)	Profit, 53%
5)	Loss, 20%	6)	Profit, 50%
7)	Profit, 22%	8)	Loss, 23%
9)	Profit, 35%	10)	Profit, 43%
11)	Loss, 63%	12)	Loss, 26%
13)	Profit, 25%	14)	Loss, 26%
15)	Profit, 46%	16)	Profit, 14%

Page 225, Earning Dividends

Objective: To compute the total dividends for an investment.

Vocabulary: Dividend ("Divide" profits at "end" of earning period).

ANSWERS:

1)	$240	2)	$7065
3)	$1257	4)	$8441.06
5)	$196.56	6)	$1228.80
7)	$124	8)	$2557.86
9)	$1755.40	10)	$10,690.56
11)	33 shares	12)	8 shares
13)	34 shares		

Page 226, Chapter Review

1) $837.50
2) $105.06
3) 4.5 years
4) Three thousand, four hundred seventy-six dollars and eighty-nine cents.
5) $355.86
6) No. It is off by $1.01.
7) $51,862.50
8) 44% loss

Chapter 11, PAYING TAXES (Pages 227-242)

Page 227, The Federal Budget

Objective: To interpret a circle graph that depicts categories of the federal budget.

Vocabulary: Revenue, receipts, corporate taxes.

Notes: To have students appreciate the staggering amounts of money involved in the federal budget, ask them to write out the answers in digits, e.g., $260.408 billion = $260,408,000,000.

Page 228

ANSWERS:

	Percent	Amount in Billions
1)	43%	$260.408
2)	29%	$175.624
3)	6%	$36.336
4)	4%	$24.224
5)	9%	$54.504
6)	9%	$54.504

Page 229

		Amount in Billions
1)	34%	$224.808
2)	24%	$158.688
3)	10%	$66.12
4)	10%	$66.12
5)	4%	$26.448
6)	4%	$26.448
7)	5%	$33.06
8)	9%	$59.508

Use the amounts given in the first sentence on each page. $55,600,000,000 more was spent than was received. ($55.6 billion)

Page 230, Reading the Tax Table

Objective: To read the tax table to determine the tax due.

Notes: Be careful about amounts that equal an upper limit of a range. The labels for the ranges state "At least" and "But **less than**." The tax for the highest income in a range will be found on the next line below.

Page 231
ANSWERS:

1) $3,965		11) $4,066
2) $4,267		12) $5,230
3) $3,446		13) $3,179
4) $4,133		14) $3,978
5) $5,357		15) $3,999
6) $3,550		16) $3,214
7) $4,368		17) $3,611
8) $2,954		18) $4,301
9) $4,055		19) $5,102
10) $4,720		20) $3,048

Page 232, Using a Tax Schedule
Objective: To compute tax using a tax schedule.

Vocabulary: Tax bracket, accompanies, joint return.

Page 233
ANSWERS:

1) $84	2) $3,933
3) $6,025.89	4) $2,017.86
5) $0	6) $27,859.00
7) $141.724.00	8) $8,180.49
9) $3,400.94	10) $7,268.46

Page 234, Refund or Balance Due
Objective: To compute the balance due or the amount to be refunded.

Page 235
ANSWERS:

1) $35 refund	11) $966 due
2) $733 refund	12) $1,107 due
3) $54 refund	13) $289 refund
4) $133 due	14) $974 due
5) $657 due	15) $21 refund
6) $320 due	16) $450 due
7) $1,268 due	17) $188 due
8) $1,346 refund	18) $386 refund
9) $46 refund	19) $1,279 due
10) $280 refund	20) $551 refund

Page 236, The Key to Ordering Decimals
Objective: To order decimals from largest to smallest and from smallest to largest.

Note: Emphasize the parallel skill of alphabetizing.

Page 237
ANSWERS:

1)	2.4	2.417	2.425	2.43	2.491
2)	7	7.003	7.03	7.19	7.216
3)	11.123	11.132	11.213	11.231	11.321
4)	.779	.7799	.77999	.78	.781
5)	.9	1	1.001	1.01	1.1
6)	5.31	5.309	5.2	5.177	5.00009
7)	2.45	2.36	2.359	2.358	2.1788
8)	4.2555	4.255	4.25	4.249	4.24
9)	11	10	10.1	10.01	10.011
10)	.1111	.111	.1101	.11	.1

Page 238, Property Tax
Objective: To compute property taxes.

Vocabulary: Assessed value, market value.

Notes: Discuss why tax is calculated on the assessed value rather than the market value of a property.

Page 239
ANSWERS:

1) $27,000		6) $42,900
2) $35,000		7) $33,700.80
3) $30,000		8) $34,801.25
4) $33,800		9) $25,802.15
5) $31,607.50		10) $37,643.04
11) 3.67%		16) 5.009%
12) 4.21%		17) 4.5%
13) 3.52%		18) 5%
14) 4.01%		19) 3.99%
15) 3.751%		20) 4.11%
21) $990.90		22) $1,473.50
23) $1,056		24) $1,355.38
25) $1,185.26		26) $2,149.29
27) $1,516.50		28) $1,740.05
29) $1,029.50		30) $1,547.13

Page 240, Effective Tax Rate
Objective: To compute the effective tax rate.

Notes: The effective tax rate is the simple percent of the property tax value which yields the same tax as the tax rate of the assessed value.

The advantage of the effective tax rates is that they are easier to compare. When deciding which subdivision has lower taxes, the effective tax rates should be compared.

ANSWERS: 1) $6,739.92 2) $6,750

Page 241

ANSWERS:
1. a) 0.012845
   b) 0.0125   b is lower.
2. a) 0.01216
   b) 0.01141   b is lower.
3. a) 0.0118075
   b) 0.012648   a is lower.
4. a) 0.01554
   b) 0.01554   a and b are the same.
5. a) 0.014025
   b) 0.01443   a is lower.
6. a) 0.00672
   b) 0.00792   a is lower.

7. a) 0.0105
   b) 0.010045   b is lower.
8. a) 0.0205
   b) 0.01961   b is lower.
9. a) 0.01476
   b) 0.01575   a is lower.
10. a) 0.00688
    b) 0.007285   a is lower.

Page 242, Chapter Review

1. $79,686,000,000
2) $4,869
3) $4,024
4) $1,944 balance due
5) $22,848.88
6) Joint, $2,795.40
   Separate, $4,343.30
7) $2,959.09
8) 4.275%
9) The market value is the amount of money that similar houses are selling for.

   The assessed value is a percentage of the market value. The percent is fixed by the assessing agency.
10) 1.83%

Chapter 12, PREPARING FOR CAREERS (Pages 243-262)

Page 243, Salesclerks

Objective: To calculate sales tax.

ANSWERS:
1) $.57   2) $1.25   3) $2.65
4) $.50   5) $1.77   6) $8.43

Page 244

Objective: To read a sales tax table.

ANSWERS:

	Tax	Total		Tax	Total
1)	$.08	$1.51	2)	$.57	$11.87
3)	$.60	$12.50	4)	$.54	$11.29
5)	$.05	$.90	6)	$1.08	$22.58
7)	$.55	$11.45	8)	$.53	$10.98
9)	$.06	$1.26	10)	$1.05	$22.05
11)	$1.05	$21.95	12)	$.55	$11.55
13)	$1.07	$22.42	14)	$.51	$10.68
15)	$1.02	$21.41	16)	$.02	$.40
17)	$1.06	$2213	18)	$.52	$10.91
19)	$1.05	$22.02	20)	$.56	$11.64

Page 245, Giving Change

Objective: To determine the correct amount of change in bills and coins.

ANSWERS:

Change Due	Bills				Coins			
	$20	$10	$5	$1	Quarters	Dimes	Nickels	Pennies
1) $16.47		1	1	1	1	2		2
2) 42¢					1	1	1	2
3) 89¢					3	1		4
4) $1.23				1		2		3
5) $3.62				3	2	1		2
6) $8.37			1	3	1	1		2
7) $11.01		1		1				1
8) $19.99		1	1	4	3	2		4
9) $24.32	1			4	1		1	2
10) $51.48	2	1		1	1	2		3
11) $78.43	3	1	1	3	1	1	1	3

Page 246, Surveyors

Objective: To apply the formula c = b tan a to physical situations.

Notes: Obtain a surveyor's transit and give students the opportunity to measure angles to solve problems that you or the class devise. This is an enjoyable outdoors activity.

Page 247

ANSWERS:
1) 24 feet
2) 5 feet
3) 100 feet
4) 20 feet
5) 46 feet

Page 248, The Key to Square Root

Objective: To compute square roots.

Notes: There are a number of methods for computing square roots. This method best makes use of the definition, i.e., equal factors.
    Try the example with the first trial divisor of 4. Show how results are independent of the first choice.

ANSWERS:
1) 2   2) 5
3) 6   4) 9
5) 10   6) 1
7) 8   8) 12
9) 7   10) 3

Page 249

1) 17.3   2) 6.2
3) 25.6   4) 11.3
5) 16   6) 4.6
7) 20.2   8) 18.6
9) 26.8   10) 32

Page 250, Electricians

Objective: To apply the electrical formulas.

Notes: Students must not be required to memorize these formulas.

Page 253:

Notes: Any number is useful for a guess in Step 3.

ANSWERS:
1) I = 8   E = 16
2) R = 4   E = 20
3) W = 840   R = 17.1
4) I = 7.28   R = 137.36
5) W = 1600   E = 160
6) I = .5   R = 220
7) W = 28800   I = 120
8) W = 792   R = 15.28
9) I = 2.2   E = 220
10) W = 2400   I = 20
11) R = 6   E = 120
12) W = 1800   E = 120

Page 254, Auto Mechanics

Objective: To order fractions.

Page 255

ANSWERS:

1) $\frac{5}{16}$ $\frac{3}{8}$ $\frac{7}{16}$ $\frac{1}{2}$ $\frac{5}{8}$

2) $1\frac{1}{8}$ $1\frac{3}{16}$ $1\frac{7}{16}$ $1\frac{1}{2}$ $1\frac{5}{8}$

3) $2\frac{9}{16}$ $2\frac{5}{8}$ $2\frac{11}{16}$ $2\frac{3}{4}$ $2\frac{7}{8}$

4) $\frac{9}{16}$ $\frac{5}{8}$ $\frac{11}{16}$ $\frac{3}{4}$ $\frac{13}{16}$

5) 1 $1\frac{1}{32}$ $1\frac{1}{16}$ $1\frac{1}{8}$ $1\frac{1}{4}$

Page 256, Carpenters

Objective: To measure line segments to varying degrees of precision.

Page 257

Notes: Students will find that for these line segments, the $\frac{1}{16}$ " measurements are equivalent to the $\frac{1}{8}$ " measurements.

ANSWERS:
1) $1\frac{1}{4}$, $1\frac{1}{8}$, $1\frac{2}{16}$

2) $2\frac{2}{4}$, $2\frac{3}{8}$, $2\frac{6}{16}$

3) $1\frac{3}{4}$, $1\frac{5}{8}$, $1\frac{10}{16}$

4) $3\frac{2}{4}$, . $3\frac{3}{8}$, $3\frac{6}{16}$

5) $1\frac{2}{4}$, $1\frac{4}{8}$, $3\frac{8}{16}$

6) $\frac{2}{4}$, $\quad\frac{3}{8}$, $\quad\frac{6}{16}$

7) $1\frac{4}{4}$, $\quad1\frac{7}{8}$, $\quad1\frac{14}{16}$

8) $\frac{1}{4}$, $\quad\frac{1}{8}$, $\quad\frac{2}{16}$

9) $\frac{4}{4}$, $\quad\frac{8}{8}$, $\quad\frac{16}{16}$

10) $3\frac{1}{4}$, $\quad3\frac{1}{8}$, $\quad3\frac{2}{16}$

11) $2\frac{1}{4}$, $\quad2\frac{1}{8}$, $\quad2\frac{2}{16}$

12) $3\frac{3}{4}$, $\quad3\frac{6}{8}$, $\quad3\frac{12}{16}$

13) $2\frac{1}{4}$, $\quad2\frac{2}{8}$, $\quad2\frac{4}{16}$

14) $2\frac{4}{4}$, $\quad2\frac{7}{8}$, $\quad2\frac{14}{16}$

15) $\frac{2}{4}$, $\quad\frac{4}{8}$, $\quad\frac{8}{16}$

16) $2\frac{2}{4}$, $\quad2\frac{4}{8}$, $\quad2\frac{8}{16}$

17) $3\frac{1}{4}$, $\quad3\frac{2}{8}$, $\quad3\frac{4}{16}$

18) $2\frac{3}{4}$, $\quad2\frac{6}{8}$, $\quad2\frac{12}{16}$

19) $3\frac{4}{4}$, $\quad3\frac{7}{8}$, $\quad3\frac{14}{16}$

20) $4\frac{1}{4}$, $\quad4\frac{1}{8}$, $\quad4\frac{2}{16}$

Page 258, Drafters

Objective: To compute the scale length for drawings.

Vocabulary: Scale, blueprints.

Page 259
ANSWERS:

1) 4
2) 12
3) 12
4) 6
5) 9
6) 49
7) 5
8) 15
9) 9
10) 40
11) $\frac{7}{16}$
12) $\frac{7}{8}$
13) $\frac{4}{7}$
14) $3\frac{3}{4}$

15) $2\frac{2}{5}$
16) 2
17) 5
18) 6
19) 7
20) $10\frac{1}{2}$

Page 260, Machine Operators

Objective: To apply the rule for RPM plus number of teeth to problems about gears.

Vocabulary: Driver, driven gear, RPM, rpm.

Notes: Point out that capital and lower case letters are very distinct in the rule. Care needs to be taken to remain alert of this.

Page 261
ANSWERS:
1) 10 teeth
2) 6 RPM
3) 7 Teeth
4) 84 rpm
5) 10 Teeth
6) 9 teeth
7) 80 RPM
8) 5 teeth
9) 30 Teeth
10) $13\frac{1}{3}$ rpm

Page 262, Chapter Review
1) 8.25 miles
2) $1.00, $20.95 total
3) One five-dollar bill, two one-dollar bills, two quarters, one dime, one nickel, and three pennies.
4) 19.1
5) 336 W
6) $10\frac{1}{16}$, $10\frac{1}{8}$, $10\frac{3}{16}$, $10\frac{1}{4}$, $10\frac{3}{8}$, $10\frac{3}{4}$
7) 2″
8) $\frac{7}{8}$ ″
9) $1\frac{1}{2}$ ″
10) 175 RPM

ACTIVITY WORKSHEETS

AND

ALTERNATE CHAPTER TESTS

• Reproducible Pages •

ACTIVITY WORKSHEETS
AND CHAPTER TESTS

These worksheets may be reproduced by the teacher and used with students to enhance their instructional program.

SURVEY TEST FOR WHOLE NUMBERS

▶ Place Value. Write the place name for each underlined digit.

1) 3<u>0</u>5 _____

2) 3,<u>9</u>13 _____

3) <u>9</u>,039 _____

4) <u>4</u>,958,509 _____

▶ Write these numerals in words.

5) 52,609 _____

6) 2,582,844 _____

_____

▶ Round these whole numbers to the nearest:

       Ten: Hundred: Thousand:

7) 469 _____

8) 2,475,521 _____

9) 489 _____

▶ Perform the indicated operations.

10)   3841   11)   9681   12)   203   13) 1024 × 10 = _____
        1382   − 773   × 36
    + 800

14) 34 + 704 + 331 + 1002 = _____

15) 50231 − 9437 = _____

16) 5)‾7551

17) 46)‾4729

18) 32048 ÷ 16 = _____

▶ Round the answer to the nearest whole number.

19) 346 ÷ 21 = _____

20) 24 − 8 × 4 ÷ 2 + 2 = _____

Name _____

Class _____ Date _____

ADDITION OF WHOLE NUMBERS

Example: $234 + 349 + 1603 =$ _____   *Write this:*

$$\left.\begin{array}{r} 234 \\ 349 \\ + \ 1603 \end{array}\right\} \leftarrow \text{Addends}$$

$$\overline{2186} \leftarrow \text{Sum}$$

▶ Add.

1) $23 + 48 + 506 =$ _____

2) $5 + 94 + 80 + 3 =$ _____

3) $5 + 74 + 102 + 49 =$ _____

4) $203 + 448 + 509 =$ _____

5) $429 + 747 + 67 =$ _____

6) $690 + 38 + 441 + 8 =$ _____

7) $3594 + 37 + 380 =$ _____

8) $405 + 393 + 4488 =$ _____

9) $48,204 + 74,503 + 302 =$ _____

10) $7834 + 6539 + 92,389 =$ _____

11) $283 + 7485 + 3774 =$ _____

12) $560 + 374 + 6005 =$ _____

13) $4759 + 5768 + 30,481 =$ _____

14) $49,036 + 87,630 + 390,476 =$ _____

15) $30,457 + 58,604 + 80,512 =$ _____

16) $38,405 + 50,067 + 40,584 =$ _____

17) $203,753 + 859,302 + 702,641 =$ _____

18) $5,037,583 + 7,458,324 + 37,495,621 =$

_____

Name _____

Class _____ Date _____

SUBTRACTION OF WHOLE NUMBERS

Example: 30045 − 4857 = _____

Write this: 30,045 ⟵ Minuend
 − 4,857 ⟵ Subtrahend
 ‾‾‾‾‾‾‾‾
 25,188 ⟵ Difference or
   Remainder

▶ Subtract.

1) 372 − 45 = _____

2) 754 − 586 = _____

3) 3841 − 548 = _____

4) 9004 − 486 = _____

5) 3945 − 459 = _____

6) 5108 − 4960 = _____

7) 4056 − 3506 = _____

8) 50,400 − 38,404 = _____

9) 30,451 − 5968 = _____

10) 84,452 − 9574 = _____

11) 95,068 − 49,052 = _____

12) 89,502 − 9495 = _____

13) 490,683 − 39,475 = _____

14) 10,237 − 9340 = _____

15) 102,873 − 30,475 = _____

16) 340,900 − 12,009 = _____

17) 340,581 − 93,364 = _____

18) 200,319 − 92,338 = _____

19) 300,471 − 49,284 = _____

20) 100,000 − 20,594 = _____

21) 101,039 − 39,900 = _____

22) 944,032 − 94,475 = _____

23) 200,341 − 90,943 = _____

24) 837,529 − 84,750 = _____

25) 838,470 − 84,757 = _____

26) 580,334 − 84,758 = _____

27) 405,562 − 393,400 = _____

28) 344,163 − 12,419 = _____

Name _____

Class _____ Date _____

MULTIPLICATION OF WHOLE NUMBERS

Example: 273 × 49 =

Write this:
$$\begin{array}{r} 273 \\ \times\ \ 49 \\ \hline 2457 \\ 1092 \\ \hline 13377 \end{array}$$
— Factors
— Partial Products
— Product

▶ Multiply.

1) 239
   × 8

2) 203
   × 12

3) 485
   × 31

4) 349
   × 11

5) 803
   × 72

6) 402
   × 84

7) 847
   × 94

8) 475
   × 47

9) 485
   × 51

10) 3102
   × 18

11) 4852
   × 27

12) 9050
   × 91

13) 2961
   × 102

14) 5706
   × 313

15) 4066
   × 116

▶ Write these in the vertical form and multiply.

16) 3041 × 325 =

17) 4712 × 482 =

18) 3012 × 384 =

Name _____

Class _____ Date _____

DIVISION OF WHOLE NUMBERS WITHOUT REMAINDERS

Example: $1404 \div 6 =$
Write this:

```
        234
    6) 1404      Quotient
       12
       20
       18          Dividend
       24
       24
```
Divisor

Example: $3120 \div 12 =$
Write this:

```
        260
    12) 3120
        24
        72
        72
```

Example: $11707 \div 23 =$
Write this:

```
         509
    23) 11707
        115
        207
        207
```

▶ Divide.

1) $9\overline{)2439}$

2) $7\overline{)3164}$

3) $8\overline{)1648}$

4) $7\overline{)2737}$

5) $8\overline{)4984}$

6) $13\overline{)6799}$

7) $13\overline{)3965}$

8) $17\overline{)1802}$

9) $29\overline{)17835}$

10) $23\overline{)16330}$

11) $31\overline{)17484}$

12) $28\overline{)61880}$

▶ Write these in the standard form and divide.

13) $4173 \div 39 =$

14) $8316 \div 18 =$

15) $5136 \div 48 =$

Name _____

Class _____ Date _____

DIVISION OF WHOLE NUMBERS WITH REMAINDERS

Example: 3259 ÷ 9 =   *Example:* 7006 ÷ 17 =   *Example:* 7543 ÷ 26 =

Write   *Write*   *Write*
this: $362\frac{1}{9}$ *this:* $412\frac{2}{17}$ *this:* $290\frac{3}{26}$

$$9\overline{)3259}$$ $$17\overline{)7006}$$ $$26\overline{)7543}$$
$$\underline{27}$$ $$\underline{68}$$ $$\underline{52}$$
$$55$$ $$20$$ $$234$$
$$\underline{54}$$ $$\underline{17}$$ $$\underline{234}$$
$$19$$ $$36$$ $$3$$
$$\underline{18}$$ $$\underline{34}$$
$$1$$ $$2$$

Remember to write the remainder over the divisor.

_____

▶ Divide.

1) $8\overline{)2345}$ 2) $7\overline{)3559}$ 3) $11\overline{)3855}$ 4) $9\overline{)5999}$

5) $18\overline{)5565}$ 6) $31\overline{)11254}$ 7) $11\overline{)3735}$ 8) $20\overline{)110019}$

9) $41\overline{)4800}$ 10) $42\overline{)4499}$ 11) $61\overline{)25499}$ 12) $40\overline{)14415}$

▶ Write these in the standard form and divide.

13) 14472 ÷ 91 =   14) 53408 ÷ 51 =   15) 72420 ÷ 65 =

REVIEW OF BASIC OPERATIONS
WITH WHOLE NUMBERS

1) $25 + 341 =$

2) $304 \times 23 =$

3) $1{,}002 - 384 =$

4) $26{,}261 \div 25 =$

5) $3{,}020 \times 105 =$

6) $80{,}345 - 2{,}934 =$

7) $7{,}022 \div 68 =$

8) $8{,}054 \times 112 =$

9) $55{,}067 + 399 + 944 =$

10) $49{,}322 \div 33 =$

11) $49{,}338 - 9{,}442 =$

12) $38 + 12 - 19 =$

13) $9{,}122 \div 8 =$

14) $30{,}091 - 28{,}949 =$

15) $7{,}456 - 234 + 283 =$

16) $801 \times 20 \div 10 =$

17) $288 + 942 + 9{,}511 =$

18) $40{,}013 - 23{,}471 =$

19) $674 + 85 - 495 =$

20) $98{,}003 - 83{,}741 =$

21) $40{,}591 \div 3 =$

22) $5{,}900 \times 400 =$

23) $10{,}384 \times 200 =$

24) $40{,}513 \div 39 =$

25) $8{,}371 - 578 =$

26) $56{,}571 \div 65 =$

27) $4 + 23 + 405 + 933 =$

28) $37 \times 14 \times 35 =$

29) $9{,}832 + 293 + 39{,}441 =$

30) $5{,}761 + 384 - 481 =$

31) $304 + 35 - 27 + 83 =$

32) $144{,}144 \div 36 =$

33) $80{,}028 - 29{,}388 =$

34) $60{,}021 \times 847 =$

35) $102{,}283 - 23{,}384 =$

36) $302 \times 21 \div 9 =$

37) $11{,}028 - 983 =$

38) $42 + 6 + 81 + 923 =$

39) $90{,}000 \div 100 =$

40) $499 + 76 + 22 - 274 =$

41) $30{,}022 \div 29 =$

42) $5{,}058 \times 501 =$

43) $58{,}007 \div 12 =$

44) $40{,}596 + 293 + 948 =$

45) $40{,}591 - 2{,}935 + 47{,}501 =$

46) $846{,}102 \div 14 =$

47) $875 + 4{,}059 + 2{,}374 =$

48) $60{,}900 \div 5{,}002 =$

SURVEY TEST FOR DECIMALS

▶ Place Value. Write the place name for each underlined digit.

1) 84.03<u>4</u> _____

2) 0.509<u>1</u>1 _____

3) 6.3<u>0</u>0499 _____

4) 293.<u>1</u>93 _____

▶ Write these numerals in words.

5) 34.072 _____

6) 0.10853 _____

_____

▶ Round these decimals to the nearest:

Tenth: Hundredth: Thousandth:

7) 4.0481 _____

8) 46.1482 _____

9) 0.09 _____

▶ Perform the indicated operations.

10) 73.407
     5.9
     0.4921
+ 103.93

11) 29.8
− 7.831

12) 2.38
× 2.4

13) 18.04 + 0.0942 + 5 + 1.1 =

14) 57.3 − 0.947 =

15) $26\overline{)48.1}$

16) $.08\overline{).0424}$

17) 0.0819 ÷ 1.3 =

▶ Round the answer to the nearest:

Tenth: Hundredth: Thousandth:

18) 7 ÷ 8 = _____

19) 1.1 ÷ 8 = _____

20) 2 ÷ 3 = _____

Name _____

Class _____ Date _____

ADDITION OF DECIMALS

Example: 3 + 2.4 + 0.06 =

Write	3		3.00
this:	2.4	OR	2.40
	+ .06		+ 0.06
	5.46		5.46

Example: 4 + 0.35 + 1.082 =

Write	4.000
this:	0.350
	+ 1.082
	5.432

Helpful Hints

a) Remember that the number 3 can be expressed as a decimal, that is, 3 = 3.0 = 3.00.

b) Remember that the decimal points must be lined up before you begin to add.

c) Remember to place the decimal point in the sum as shown in the example.

d) Remember to place zeros in the addends to help with the addition.

▶ Add. Place zeros in the addends.

1)	3.00	2)	4.00	3)	5.09	4)	1.026
	2.93		5.103		2.036		4.56
	+ .78		23.049		+ 90.345		63.0071
			+ 2.9012				+ 1.

5)	34.03	6)	6.7	7)	.506	8)	6.3
	5.602		.347		41.0033		.037
	3.8401		9.62		9.1		7.0322
	+ 23.1		+ 2.2		+ 61		+ 82.9

9)	39.041	10)	923.1	11)	3.3	12)	402.1005
	6.7		73.12		.0093		61.03
	5.06		7.00002		73.00381		4.6
	+ 74		+ .64		+ 2920.08		+ 22.37

▶ Write these in the vertical form and add.

13) 2.3 + 0.46 + 91.308 = _____   14) 8 + 3.9 + 0.73 = _____

15) 7.5 + 4.4 + 5 = _____   16) 0.76 + 1.3 + 6 = _____

17) 5.8 + 1 + 0.406 = _____   18) 66.02 + 8.1 + 5 = _____

19) 8 + .702 + 32.1 = _____   20) 63 + 4.56 + 5.8 = _____

Name _____

Class _____ Date _____

SUBTRACTION OF DECIMALS

Example: 3.63 − 0.734 =

Write   3.630 ⟵ Insert a zero
this: − .734   here.
            ‾‾‾‾‾‾‾
             2.896

Helpful Hints

a) Remember to fill places in the minuend with zeros.

Example: 8 − 0.631 =

Write   8.000 ⟵ Insert zeros
this: − .631   here.
            ‾‾‾‾‾‾‾
             7.369

b) Remember to keep the decimal points lined up.

▶ Insert zeros and subtract.

1) 34.3 − 5.64	2) 4 − .349	3) 7.302 − .83	4) 5.1 −1.204

5) 48.22 − 3.489	6) 39.4 − .0371	7) 10 − 3.4005	8) 356.748 − 7.8

9) 5.602 − 4.0498	10) 81.923 − 23.9047	11) 38. − .0273	12) 9 − .9

13) 3 − .0234	14) 74.73 − 5.332	15) 7465.2 − .9098	16) 37 − 8.394

▶ Write these in the vertical form and subtract.

17) 23.4 − 4.56 = _____   18) 4 − 0.48 = _____

19) 63.2 − 4.509 = _____   20) 16 − 1.34 = _____

21) 82 − 2.302 = _____   22) 38.809 − 7.7081 = _____

23) 9 − 3.4051 = _____   24) 0.983 − 0.01023 = _____

Name _____

Class _____ Date _____

MULTIPLICATION OF DECIMALS

Example: 31.2 × 0.34 =

Write	31.2	1 decimal place
this:	× .34	+ 2 decimal places
	1248	3 decimal places to
	936	be marked off in the
	10.608	product counting from
		right to left.

Example: 0.33 × 0.005 =

Write	.33	
this:	× .005	Sometimes it
	.00165	becomes necessary
		to insert zeros
		at the left.

► Multiply.

1) 3.4
    × 2.6

2) 71.8
    × .29

3) 3.02
    × .12

4) 4.21
    × 3.8

5) 10.8
    × 1.71

6) 4.501
    × 2.3

7) 20.34
    × 10.3

8) .234
    × .008

9) 1.03
    × .009

10) .0037
     × .019

11) .00319
     × .0084

12) .0028
     × .072

► Write these in vertical form and multiply.

13) 2.034 × 4.5 = _____

14) 4.9 × .009 = _____

15) .004 × .24 = _____

16) 49.5 × 3.4 = _____

17) 3.405 × .003 = _____

18) .00391 × .019 = _____

19) .934 × 23.1 = _____

20) .0201 × .039 = _____

21) .0031 × .009 = _____

22) 10.07 × .35 = _____

23) 129 × 4.03 = _____

24) .506 × .0001 = _____

Name _____

Class _____ Date _____

DIVISION OF DECIMALS

Example: 18.4 ÷ 8 =

Write this:

```
        2.3 ← Quotient
    8) 18.4
       16   ← Dividend
        2 4
        2 4
```

Divisor

Example: 0.768 ÷ 1.6 =

Write this:

```
         .48
  1.6) .768
       64
       128
       128
```

Steps to Remember

a) Move the decimal point in the divisor to the right.

b) Move the decimal point in the dividend the same number of places.

c) Then place a decimal point straight above it in the quotient.

▶ Divide.

1) 9) 41.4

2) 5) 3.65

3) 8) 41.6

4) 15) 40.5

5) 1.9) 8.74

6) 1.1) 84.7

7) .72) 1.224

8) 4.3) 8.213

9) .026) 1.352

10) .57) .9234

11) .33) 6.27

12) .056) .6328

13) .26) .10946

14) .09) .1107

15) .36) 32.76

16) .04) .844

▶ Write these in the standard form and divide.

17) 0.06734 ÷ 0.037 = _____

18) 6.592 ÷ 0.16 = _____

19) 0.08357 ÷ 0.61 = _____

20) 0.4212 ÷ 0.36 = _____

21) 0.0405 ÷ 0.015 = _____

22) 0.04592 ÷ 0.41 = _____

Name _____

Class _____ Date _____

ZEROS IN THE QUOTIENT

Example: 0.01449 ÷ 0.23 =

Write this:
```
         .063
  .23).01449
       138
        69
        69
```

Example: 2.9484 ÷ 4.2 =

Write this:
```
        .702
  4.2)2.9484
      94
       84
       84
```

Divide.

1) 7.3)29.273

2) 5.2).3224

3) 1.5).0345

4) .07).1442

5) 6.3)6.741

6) .19).00247

7) .013).06513

8) 1.1).0308

9) .31)34.131

10) .022).26422

11) 16).624

12) .65)3.939

13) .41)13.1241

14) 35).3185

15) 5.7)6.042

16) .44)5.2844

17) 7.7)7.7077

18) .83)1.6683

19) .063)5.0463

20) .006).06618

▶ Write these in the standard form and divide.

21) 0.57252 ÷ 0.52 = _____

22) 0.06307 ÷ 0.007 = _____

23) 0.00748 ÷ 0.68 = _____

24) 0.26664 ÷ 1.32 = _____

Name _____

Class _____ Date _____

ROUNDING THE QUOTIENT

Example: Round to the nearer tenth.

$8 \div .9 =$

Write this:

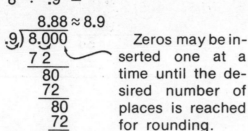

$$8.88 \approx 8.9$$

Zeros may be inserted one at a time until the desired number of places is reached for rounding.

Example: Round to the nearer hundredth.

$0.89 \div 2.3 =$

Write this:

```
         .386 ≈ .39
2.3) .8900
      69
      200
      184
      160
      138
       22
```

Reminder: It may be necessary to write zeros in the dividend.

▶ Divide. Round to the place indicated.

1) Tenth

.6)5

2) Hundredth

.06)7.1

3) Hundredth

23)1

4) Thousandth

5.8)5.9

5) Hundredth

1.9)2.5

6) Tenth

.9).87

7) Hundredth

1.3).14

8) Hundredth

.62)5.3

9) Thousandth

.03)2

10) Hundredth

6.3).64

11) Hundredth

5.1)7.2

12) One

.024)1.04

13) Hundredth

7.1)6.3

14) Tenth

9)10

15) Thousandth

6)2

16) Hundredth

12)1.45

▶ Write these in the standard form and divide. Round the quotients to the nearer hundredth.

17) $5.1 \div 7.6 =$ _____

18) $1.7 \div 0.16 =$ _____

19) $0.215 \div 0.34 =$ _____

20) $5 \div 0.32 =$ _____

60

REVIEW OF BASIC OPERATIONS
WITH DECIMALS

1) $2.3 + 5 + 0.941 =$

2) $4.5 - 0.931 =$

3) $4.5 \times 0.31 =$

4) $24.24 \div 4.8 =$

5) $7 - 0.8012 =$

6) $0.0387 \times 0.64 =$

7) $3{,}001 \times 3.4 =$

8) $3.04 - 0.95 =$

9) $6.9 \times 0.34 =$

10) $5.6 + 91 + 0.76 =$

11) $9.01 - 3.9 =$

12) $234 \div 45 =$

13) $3.4 \times 0.023 =$

14) $44.4 - 7.816 =$

15) $39.648 \div 5.6 =$

16) $801.11 - 34.551 =$

17) $18.29 \div 3.1 =$

18) $0.027 \times 0.006 =$

19) $0.603 \times 0.101 =$

20) $91 + 2.7 + 0.003 =$

21) $0.029 - 0.02001 =$

22) $74.358 \div 27 =$

23) $0.0303 \times 4.1 =$

24) $384.04 + 0.927 + 0.1 =$

25) $21.25 \div 2.5 =$

26) $3 + 4.5 + 2.21 =$

27) $3.5 - 1.29 =$

28) $83 + 2.3 + 0.939 =$

29) $1.7064 \div 0.24 =$

30) $85 + 3.53 + 2 + 0.75 =$

31) $0.1632 \div 0.08 =$

32) $9 - 0.99 =$

33) $3.096 \div 1.2 =$

34) $6.22 \times 0.002 =$

35) $2 - 1.402 =$

36) $401.1 + 29.53 + 1.2 =$

37) $61 - 0.28 =$

38) $0.4387 - 0.41 =$

39) $7 + 2.8 + 1 + 34.99 =$

40) $49.01 + 3 + 2.31 + 8 =$

41) $58.3 \times 2.4 =$

42) $4.1 - 3.009 =$

43) $0.00528 \div 0.66 =$

44) $0.08917 - 0.00991 =$

45) $73.94 + 5.6 + 2 + 0.916 =$

46) $1 - 0.1028 =$

47) $804.04 - 290.192 =$

48) $5.20251 \div 5.1 =$

SURVEY TEST FOR FRACTIONS

▶ Compare the fractions in each pair.   Use < or >.

1) $\frac{4}{9}$ $\frac{6}{11}$

2) $\frac{10}{21}$ $\frac{2}{7}$

3) $\frac{7}{24}$ $\frac{11}{42}$

▶ Express each fraction in lowest terms.

4) $\frac{9}{24}$ =

5) $\frac{18}{54}$ =

6) $14\frac{9}{33}$ =

7) $\frac{64}{120}$ =

▶ Rename each mixed number as an improper fraction.

8) $6\frac{2}{7}$ =

9) $3\frac{5}{9}$ =

10) $13\frac{1}{8}$ =

11) $4\frac{7}{12}$ =

▶ Rename each improper fraction as a mixed number.

12) $\frac{34}{11}$ =

13) $\frac{63}{9}$ =

14) $\frac{85}{4}$ =

15) $\frac{23}{3}$ =

▶ Perform the indicated operations.

16) $\frac{1}{7} \times \frac{2}{9}$ =

17) $1\frac{5}{16} \times 2\frac{8}{9}$ =

18) $2\frac{2}{9} \times 10$ =

19) $\frac{7}{8} \div \frac{9}{16}$ =

20) $3\frac{2}{5} \div 3\frac{1}{11}$ =

21) $\frac{12}{13} \div 24$ =

22) $7\frac{2}{5}$
  $+ \ \frac{4}{5}$
  _____

23) $12\frac{1}{8}$
  $+ \ 25\frac{1}{6}$
  _____

24) $31\frac{7}{15}$
  $+ \ 8\frac{1}{3}$
  _____

25) $13 - 2\frac{5}{8}$ =

26) $53\frac{3}{14} - 19\frac{6}{7}$ =

Name _____

Class _____ Date _____

EXPRESSING FRACTIONS IN HIGHER TERMS

Example: Express $\frac{5}{6}$ as a fraction with a denominator of 24.

Step 1:	Step 2:	Step 3:	Step 4:
$\frac{5}{6} = \frac{}{24}$	$\frac{5 \times 4}{6 \times 4} = \frac{}{24}$	$\frac{5 \times 4}{6 \times 4} = \frac{20}{24}$	$\frac{5}{6} = \frac{20}{24}$

Because $24 \div 6 = 4$, multiply 5 by 4.　　　New Fraction

► Express each fraction in higher terms as indicated.

1) $\frac{7}{8} = \frac{}{40}$　　2) $\frac{4}{9} = \frac{}{36}$　　3) $\frac{2}{3} = \frac{}{12}$　　4) $\frac{5}{11} = \frac{}{55}$　　5) $\frac{5}{12} = \frac{}{36}$

6) $\frac{2}{7} = \frac{}{35}$　　7) $\frac{6}{9} = \frac{}{54}$　　8) $\frac{1}{2} = \frac{}{10}$　　9) $\frac{5}{13} = \frac{}{39}$　　10) $\frac{4}{15} = \frac{}{75}$

11) $\frac{3}{11} = \frac{}{66}$　　12) $\frac{2}{17} = \frac{}{34}$　　13) $\frac{12}{20} = \frac{}{60}$　　14) $\frac{11}{12} = \frac{}{60}$　　15) $\frac{4}{21} = \frac{}{84}$

16) $\frac{1}{16} = \frac{}{48}$　　17) $\frac{3}{13} = \frac{}{65}$　　18) $\frac{4}{22} = \frac{}{110}$　　19) $\frac{5}{7} = \frac{}{56}$　　20) $\frac{3}{5} = \frac{}{95}$

21) $\frac{3}{9} = \frac{}{54}$　　22) $\frac{1}{7} = \frac{}{63}$　　23) $\frac{2}{3} = \frac{}{108}$　　24) $\frac{3}{4} = \frac{}{52}$　　25) $\frac{12}{21} = \frac{}{126}$

26) $\frac{2}{11} = \frac{}{121}$　　27) $\frac{3}{16} = \frac{}{80}$　　28) $\frac{4}{5} = \frac{}{80}$　　29) $\frac{2}{12} = \frac{}{84}$　　30) $\frac{5}{7} = \frac{}{70}$

31) $\frac{2}{12} = \frac{}{72}$　　32) $\frac{3}{18} = \frac{}{54}$　　33) $\frac{5}{16} = \frac{}{112}$　　34) $\frac{2}{19} = \frac{}{76}$　　35) $\frac{5}{13} = \frac{}{91}$

36) $\frac{6}{15} = \frac{}{105}$　　37) $\frac{4}{13} = \frac{}{117}$　　38) $\frac{11}{23} = \frac{}{161}$　　39) $\frac{35}{50} = \frac{}{250}$　　40) $\frac{5}{40} = \frac{}{200}$

Name _____

Class _____ Date _____

RENAMING TO LOWEST TERMS

Example: $\frac{12}{16} = \frac{12 \div 4}{16 \div 4} = \frac{3}{4}$

Divide the numerator and the denominator by 4 because 4 is a common factor of 12 and 16.

Example: $3\frac{12}{16} = 3 + \frac{12}{16} = 3 + \frac{3}{4} = 3\frac{3}{4}$

Rename $\frac{12}{16}$ as shown in the first example.

▶ Rename each fraction to the lowest terms.

1) $\frac{8}{10} =$

2) $10\frac{10}{12} =$

3) $\frac{12}{36} =$

4) $\frac{16}{18} =$

5) $\frac{22}{44} =$

6) $\frac{10}{16} =$

7) $3\frac{5}{25} =$

8) $8\frac{14}{22} =$

9) $\frac{28}{38} =$

10) $\frac{26}{36} =$

11) $\frac{52}{64} =$

12) $\frac{13}{65} =$

13) $\frac{9}{21} =$

14) $2\frac{20}{42} =$

15) $\frac{16}{24} =$

16) $\frac{15}{21} =$

17) $\frac{42}{57} =$

18) $6\frac{14}{35} =$

19) $17\frac{30}{54} =$

20) $11\frac{2}{10} =$

21) $\frac{18}{81} =$

22) $\frac{27}{81} =$

23) $\frac{40}{56} =$

24) $\frac{24}{56} =$

25) $\frac{14}{18} =$

26) $\frac{28}{35} =$

27) $\frac{84}{108} =$

28) $9\frac{12}{20} =$

29) $\frac{32}{48} =$

30) $\frac{56}{63} =$

31) $\frac{15}{51} =$

32) $\frac{45}{57} =$

33) $\frac{18}{63} =$

34) $\frac{48}{72} =$

35) $\frac{54}{68} =$

36) $\frac{72}{104} =$

37) $\frac{60}{84} =$

38) $\frac{39}{91} =$

39) $7\frac{28}{42} =$

40) $23\frac{76}{84} =$

Name _____

Class _____ Date _____

RENAMING TO THE SIMPLEST FORM

Example: $\frac{9}{7}$

Think: $7 \overline{)\begin{array}{c} 1 \\ 9 \\ 7 \\ \hline 2 \end{array}}$ equals $1\frac{2}{7}$

Answer: $\frac{9}{7} = 1\frac{2}{7}$

Example: $16\frac{15}{4}$

$16 + \frac{15}{4}$

$16 + 3\frac{3}{4}$

$19\frac{3}{4}$

$4\overline{)\begin{array}{c} 3 \\ 15 \\ 12 \\ \hline 3 \end{array}}$ equals $3\frac{3}{4}$

▶ Rename each to the simplest form.

1) $\frac{18}{5} =$ 　　2) $16\frac{4}{3} =$ 　　3) $\frac{19}{2} =$ 　　4) $\frac{22}{7} =$ 　　5) $\frac{25}{3} =$

6) $\frac{28}{5} =$ 　　7) $\frac{23}{5} =$ 　　8) $\frac{22}{4} =$ 　　9) $23\frac{16}{9} =$ 　　10) $\frac{19}{6} =$

11) $\frac{42}{5} =$ 　　12) $\frac{35}{8} =$ 　　13) $\frac{26}{13} =$ 　　14) $\frac{32}{7} =$ 　　15) $25\frac{5}{4} =$

16) $\frac{33}{10} =$ 　　17) $13\frac{5}{2} =$ 　　18) $\frac{29}{7} =$ 　　19) $\frac{57}{6} =$ 　　20) $\frac{64}{7} =$

21) $\frac{108}{9} =$ 　　22) $\frac{123}{11} =$ 　　23) $\frac{45}{7} =$ 　　24) $33\frac{16}{3} =$ 　　25) $5\frac{18}{9} =$

26) $1\frac{32}{7} =$ 　　27) $\frac{16}{3} =$ 　　28) $\frac{47}{8} =$ 　　29) $\frac{53}{13} =$ 　　30) $2\frac{3}{2} =$

31) $\frac{53}{10} =$ 　　32) $\frac{75}{8} =$ 　　33) $6\frac{5}{4} =$ 　　34) $7\frac{4}{3} =$ 　　35) $9\frac{21}{4} =$

Name _____

Class _____ Date _____

ADDITION OF FRACTIONS

Example: $12\frac{1}{5} + 4\frac{3}{5} =$

Write this:
$$\begin{array}{r} 12\frac{1}{5} \\ + 4\frac{3}{5} \\ \hline 16\frac{4}{5} \end{array}$$

If the denominators are the same, then add the numerators.

Example: $13\frac{2}{7} + 3\frac{3}{14} =$

Write this:
$$\begin{array}{r} 13\frac{2}{7} = 13\frac{4}{14} \\ + 3\frac{3}{14} = 3\frac{3}{14} \\ \hline 16\frac{7}{14} = 16\frac{1}{2} \end{array}$$

Find the least common denominator. Then add.

Simplified to the lowest terms

▶ Add. Simplify your answers to the lowest terms.

1) $13\frac{3}{8}$
    $+ 2\frac{2}{8}$

2) $23\frac{5}{17}$
    $+ 5\frac{2}{17}$

3) $18\frac{1}{2}$
    $+ 9\frac{1}{5}$

4) $5\frac{2}{13}$
    $+ 6\frac{3}{26}$

5) $3\frac{1}{7}$
    $+ 2\frac{1}{8}$

6) $8\frac{5}{12}$
    $+ \frac{1}{6}$

7) $9\frac{2}{3}$
    $+ 5$

8) $\frac{5}{6}$
    $+ \frac{1}{5}$

9) $\frac{8}{22}$
    $+ \frac{5}{22}$

10) $2\frac{3}{10}$
     $+ 1\frac{5}{20}$

11) $35\frac{6}{7}$
     $+ 4\frac{1}{8}$

12) $14\frac{2}{10}$
     $+ 3\frac{1}{5}$

13) $33\frac{5}{8}$
     $+ \frac{1}{6}$

14) $\frac{3}{15}$
     $+ \frac{2}{30}$

15) $\frac{6}{7}$
     $+ \frac{4}{8}$

16) $3\frac{3}{8}$
     $+ 2\frac{1}{6}$

17) $5\frac{1}{3}$
     $+ 2\frac{3}{5}$

18) $9\frac{1}{6}$
     $+ 2\frac{1}{9}$

19) $8\frac{2}{11}$
     $+ 5\frac{5}{66}$

20) $2\frac{1}{5}$
     $+ \frac{8}{45}$

21) $8\frac{6}{19}$
     $+ 2\frac{3}{38}$

22) $32\frac{3}{16}$
     $+ 1\frac{2}{64}$

23) $2\frac{5}{13}$
     $+ 5$

24) $21\frac{5}{7}$
     $+ 4\frac{6}{8}$

Name _____

Class _____ Date _____

SUBTRACTION OF FRACTIONS

Example: $13\frac{11}{12} - 2\frac{2}{12} =$

Write this: $\begin{array}{r} 13\frac{11}{12} \\ - 2\frac{2}{12} \\ \hline 11\frac{9}{12} = 11\frac{3}{4} \end{array}$ If the denominators are the same, then subtract the numerators.

Simplified to the lowest terms.

Example: $6\frac{5}{7} - 2\frac{3}{21} =$

Write this: $\begin{array}{r} 6\frac{5}{7} = 6\frac{15}{21} \\ - 2\frac{3}{21} = 2\frac{3}{21} \\ \hline 4\frac{12}{21} = 4\frac{4}{7} \end{array}$ Find the least common denominator. Then subtract.

▶ Subtract. Simplify your answers to the lowest terms.

1) $\begin{array}{r} \frac{6}{7} \\ - \frac{4}{7} \\ \hline \end{array}$

2) $\begin{array}{r} 14\frac{11}{15} \\ - 2\frac{1}{15} \\ \hline \end{array}$

3) $\begin{array}{r} 8\frac{2}{3} \\ - 6\frac{1}{6} \\ \hline \end{array}$

4) $\begin{array}{r} 7\frac{4}{5} \\ - 2\frac{6}{10} \\ \hline \end{array}$

5) $\begin{array}{r} 6\frac{19}{20} \\ - 4\frac{1}{5} \\ \hline \end{array}$

6) $\begin{array}{r} 25\frac{5}{7} \\ - 2\frac{3}{8} \\ \hline \end{array}$

7) $\begin{array}{r} 2\frac{2}{3} \\ - 1\frac{1}{7} \\ \hline \end{array}$

8) $\begin{array}{r} 10\frac{3}{16} \\ - 1\frac{1}{32} \\ \hline \end{array}$

9) $\begin{array}{r} 3\frac{7}{12} \\ - \frac{2}{8} \\ \hline \end{array}$

10) $\begin{array}{r} 12\frac{4}{5} \\ - 3 \\ \hline \end{array}$

11) $\begin{array}{r} 26\frac{3}{8} \\ - 4\frac{2}{6} \\ \hline \end{array}$

12) $\begin{array}{r} 2\frac{7}{11} \\ - 1\frac{6}{66} \\ \hline \end{array}$

13) $\begin{array}{r} 3\frac{5}{8} \\ - 2\frac{3}{16} \\ \hline \end{array}$

14) $\begin{array}{r} 8\frac{5}{12} \\ - 2\frac{2}{18} \\ \hline \end{array}$

15) $\begin{array}{r} 18\frac{2}{5} \\ - 3\frac{1}{15} \\ \hline \end{array}$

16) $\begin{array}{r} 7\frac{8}{9} \\ - 2\frac{3}{18} \\ \hline \end{array}$

17) $\begin{array}{r} 26\frac{7}{8} \\ - 2\frac{1}{6} \\ \hline \end{array}$

18) $\begin{array}{r} 9\frac{5}{12} \\ - 4\frac{2}{9} \\ \hline \end{array}$

19) $\begin{array}{r} 1\frac{27}{28} \\ - \frac{3}{7} \\ \hline \end{array}$

20) $\begin{array}{r} 14\frac{1}{5} \\ - 5\frac{1}{8} \\ \hline \end{array}$

21) $\begin{array}{r} 30\frac{3}{13} \\ - 4\frac{4}{39} \\ \hline \end{array}$

22) $\begin{array}{r} 15\frac{1}{2} \\ - 2\frac{3}{7} \\ \hline \end{array}$

23) $\begin{array}{r} 81\frac{2}{11} \\ - 3\frac{2}{22} \\ \hline \end{array}$

24) $\begin{array}{r} 12\frac{6}{10} \\ - 3\frac{2}{25} \\ \hline \end{array}$

Name _____

Class _____ Date _____

SUBTRACTION WITH RENAMING

Example: $10\frac{4}{11} - 4\frac{5}{11} =$

Write this:
$$10\frac{4}{11} = 9\frac{15}{11}$$
$$- 4\frac{5}{11} = 4\frac{5}{11}$$
$$\overline{ 5\frac{10}{11}}$$

Remember, $1 = \frac{11}{11}$.

So $\frac{15}{11} = \frac{4}{11} + \frac{11}{11}$.

Example: $9\frac{2}{5} - 6\frac{11}{15} =$

Write this:
$$9\frac{2}{5} = 9\frac{6}{15} = 8\frac{21}{15}$$
$$- 6\frac{11}{15} = 6\frac{11}{15} = 6\frac{11}{15}$$
$$\overline{\phantom{-6\frac{11}{15} = 6\frac{11}{15} = } 2\frac{10}{15} = 2\frac{2}{3}}$$

▶ Subtract. Rename when necessary. Simplify your answers.

1) $12\frac{5}{13}$
$- 3\frac{6}{13}$

2) $5\frac{2}{7}$
$- 3\frac{4}{7}$

3) $16\frac{2}{3}$
$- 5\frac{3}{4}$

4) $18\frac{1}{5}$
$- 2\frac{6}{7}$

5) $33\frac{15}{18}$
$- \frac{8}{9}$

6) $41\frac{7}{10}$
$- 3\frac{4}{5}$

7) $36\frac{3}{14}$
$- 4\frac{6}{7}$

8) $45\frac{1}{9}$
$- 2\frac{3}{10}$

9) $4\frac{2}{15}$
$- 2\frac{1}{5}$

10) $29\frac{3}{16}$
$- 4\frac{5}{8}$

11) $11\frac{2}{11}$
$- 3\frac{8}{22}$

12) $29\frac{6}{31}$
$- 4\frac{21}{62}$

13) $25\frac{1}{6}$
$- 2\frac{1}{4}$

14) $4\frac{1}{6}$
$- 2\frac{3}{8}$

15) $3\frac{1}{3}$
$- 2\frac{4}{5}$

16) $1\frac{7}{8}$
$- \frac{8}{9}$

17) $27\frac{5}{16}$
$- 2\frac{7}{8}$

18) $4\frac{1}{3}$
$- 2\frac{1}{2}$

19) 18
$- 2\frac{1}{5}$

20) $17\frac{2}{9}$
$- 3\frac{4}{8}$

21) 8
$- 2\frac{1}{4}$

22) $15\frac{3}{4}$
$- 5\frac{9}{10}$

23) $3\frac{2}{9}$
$- 2\frac{1}{3}$

24) $4\frac{6}{7}$
$- 3$

Name _____

Class _____ Date _____

RENAMING MIXED NUMBERS

Example: Rename $3\frac{2}{8}$ as an improper fraction.

$3\frac{2}{8} =$

$3 \times 8 + 2 = 26$ 26 is the new numerator.
Keep 8 as the denominator.

$3\frac{2}{8} = \frac{26}{8}$

▶ Rename these mixed numbers as improper fractions.

1) $2\frac{3}{4} =$ 2) $1\frac{1}{2} =$ 3) $1\frac{1}{3} =$ 4) $2\frac{5}{8} =$

5) $3\frac{2}{3} =$ 6) $5\frac{1}{6} =$ 7) $3\frac{1}{7} =$ 8) $4\frac{2}{3} =$

9) $6\frac{2}{5} =$ 10) $5\frac{1}{5} =$ 11) $4\frac{5}{6} =$ 12) $4\frac{2}{5} =$

13) $6\frac{2}{7} =$ 14) $3\frac{4}{5} =$ 15) $9\frac{4}{9} =$ 16) $7\frac{2}{7} =$

17) $2\frac{5}{9} =$ 18) $8\frac{1}{8} =$ 19) $10\frac{1}{2} =$ 20) $11\frac{2}{3} =$

21) $9\frac{1}{3} =$ 22) $16\frac{2}{3} =$ 23) $11\frac{3}{4} =$ 24) $10\frac{1}{4} =$

25) $8\frac{4}{11} =$ 26) $5\frac{7}{10} =$ 27) $25\frac{2}{3} =$ 28) $20\frac{2}{3} =$

29) $11\frac{1}{2} =$ 30) $18\frac{1}{2} =$ 31) $26\frac{1}{2} =$ 32) $18\frac{1}{2} =$

33) $20\frac{17}{20} =$ 34) $8\frac{5}{12} =$ 35) $5\frac{2}{11} =$ 36) $35\frac{2}{4} =$

37) $15\frac{2}{3} =$ 38) $32\frac{3}{4} =$ 39) $1\frac{4}{5} =$ 40) $1\frac{3}{7} =$

41) $5\frac{5}{6} =$ 42) $13\frac{1}{3} =$ 43) $12\frac{5}{12} =$ 44) $8\frac{1}{5} =$

45) $20\frac{5}{11} =$ 46) $15\frac{4}{5} =$ 47) $13\frac{2}{3} =$ 48) $8\frac{9}{10} =$

49) $17\frac{1}{2} =$ 50) $22\frac{3}{4} =$ 51) $12\frac{1}{5} =$ 52) $10\frac{3}{8} =$

MULTIPLICATION

Example: $\frac{2}{5} \times \frac{10}{13} = \frac{20}{65} = \frac{4}{13}$ $\underline{\text{numerator times numerator}}$
                                                                    denominator times denominator

OR $\frac{2}{\cancel{5}_1} \times \frac{\cancel{10}^2}{13} = \frac{4}{13}$ Because $\frac{10}{5} = \frac{2}{1}$

Example: $2\frac{1}{2} \times 2\frac{2}{3} =$

$\frac{5}{\cancel{2}_1} \times \frac{\cancel{8}^4}{3} = \frac{20}{3} = 6\frac{2}{3}$ Because $\frac{8}{2} = \frac{4}{1}$

▶ Multiply. Simplify your answers.

1) $\frac{5}{6} \times \frac{2}{3} =$

2) $\frac{4}{5} \times \frac{3}{4} =$

3) $\frac{7}{8} \times \frac{5}{14} =$

4) $\frac{3}{8} \times \frac{10}{12} =$

5) $\frac{7}{13} \times \frac{2}{7} =$

6) $\frac{3}{11} \times \frac{22}{24} =$

7) $\frac{6}{13} \times \frac{5}{12} =$

8) $\frac{9}{10} \times \frac{5}{18} =$

9) $2\frac{1}{2} \times \frac{4}{5} =$

10) $3\frac{2}{3} \times \frac{3}{5} =$

11) $\frac{2}{7} \times 3\frac{1}{4} =$

12) $\frac{4}{8} \times 1\frac{1}{9} =$

13) $1\frac{1}{5} \times 2\frac{2}{3} =$

14) $1\frac{1}{6} \times 2\frac{1}{3} =$

15) $2\frac{3}{5} \times 1\frac{2}{7} =$

16) $3\frac{2}{5} \times \frac{15}{17} =$

17) $4\frac{3}{4} \times \frac{4}{5} =$

18) $2\frac{3}{7} \times \frac{2}{17} =$

19) $5\frac{2}{7} \times \frac{1}{7} =$

20) $10\frac{2}{3} \times \frac{15}{16} =$

21) $3\frac{1}{7} \times 1\frac{7}{11} =$

22) $\frac{2}{3} \times \frac{5}{1} =$

23) $3\frac{1}{2} \times \frac{6}{1} =$

24) $4\frac{3}{5} \times 5 =$

25) $7\frac{1}{2} \times 3\frac{1}{2} =$

26) $4\frac{2}{5} \times 1\frac{2}{3} =$

27) $4\frac{1}{5} \times \frac{1}{2} =$

28) $30\frac{1}{2} \times \frac{2}{3} =$

Name _____

Class _____ Date _____

DIVISION

Example: $\frac{5}{6} \div \frac{7}{10} =$ ⌐ Divisor

Rule: Invert the divisor $\frac{7}{10}$ to $\frac{10}{7}$ and multiply.

$$\frac{5}{\cancel{6}_3} \times \frac{\cancel{10}^5}{7} = \frac{25}{21} = 1\frac{4}{21}$$

Example: $2\frac{1}{2} \div 3\frac{1}{4} =$

$$\frac{5}{2} \div \frac{13}{4} =$$

Express mixed numbers as improper fractions.

$$\frac{5}{\cancel{2}_1} \times \frac{\cancel{4}^2}{13} = \frac{10}{13}$$

Invert the divisor. Then multiply.
Simplify if possible.

▶ Divide. Simplify your answers.

1) $\frac{5}{7} \div \frac{5}{6} =$ 2) $\frac{3}{5} \div \frac{1}{5} =$ 3) $\frac{2}{7} \div \frac{4}{14} =$ 4) $\frac{5}{8} \div \frac{6}{8} =$

5) $\frac{7}{18} \div \frac{4}{9} =$ 6) $\frac{3}{13} \div \frac{2}{5} =$ 7) $\frac{8}{9} \div \frac{5}{6} =$ 8) $\frac{7}{10} \div \frac{14}{25} =$

9) $\frac{6}{9} \div \frac{12}{18} =$ 10) $\frac{2}{3} \div \frac{1}{5} =$ 11) $1\frac{8}{9} \div \frac{5}{6} =$ 12) $\frac{10}{11} \div \frac{5}{6} =$

13) $2\frac{3}{4} \div \frac{3}{8} =$ 14) $1\frac{5}{6} \div \frac{13}{12} =$ 15) $2\frac{1}{5} \div \frac{22}{10} =$ 16) $3\frac{2}{5} \div \frac{34}{15} =$

17) $1\frac{2}{7} \div 1\frac{1}{5} =$ 18) $3\frac{2}{7} \div 1\frac{2}{8} =$ 19) $\frac{5}{8} \div 2\frac{3}{7} =$ 20) $8\frac{2}{3} \div \frac{13}{18} =$

21) $5\frac{1}{3} \div 1\frac{1}{3} =$ 22) $7\frac{1}{3} \div \frac{2}{4} =$ 23) $5\frac{3}{5} \div \frac{14}{30} =$ 24) $9\frac{2}{6} \div \frac{2}{8} =$

REVIEW OF BASIC OPERATIONS
WITH FRACTIONS

1) $\frac{3}{7} \times \frac{5}{6} =$

2) $2\frac{2}{5} + 3\frac{4}{9} =$

3) $13 + \frac{13}{24} =$

4) $\frac{7}{8} + 5\frac{1}{6} =$

5) $\frac{23}{24} \times \frac{48}{46} =$

6) $12\frac{2}{9} - 2\frac{3}{27} =$

7) $\frac{6}{7} \times 6\frac{1}{6} =$

8) $\frac{12}{25} \div \frac{16}{18} =$

9) $65 - \frac{24}{25} =$

10) $\frac{11}{25} \div 2\frac{1}{5} =$

11) $8\frac{2}{5} - 5\frac{4}{5} =$

12) $5 - 2\frac{7}{15} =$

13) $6\frac{2}{11} \times 1\frac{1}{34} =$

14) $8\frac{10}{17} - 3\frac{15}{17} =$

15) $46\frac{27}{28} - 32\frac{5}{7} =$

16) $23 \div 1\frac{2}{13} =$

17) $\frac{4}{7} + \frac{3}{8} =$

18) $1\frac{1}{7} \times 2 =$

19) $2\frac{1}{13} \times \frac{5}{9} =$

20) $1\frac{6}{7} \div \frac{13}{14} =$

21) $\frac{4}{9} \div \frac{16}{72} =$

22) $\frac{5}{12} + \frac{8}{60} =$

23) $19 - 9\frac{1}{7} =$

24) $5\frac{2}{3} + 12\frac{6}{10} =$

25) $18 - 2\frac{1}{24} =$

26) $12\frac{10}{11} - 10\frac{21}{22} =$

27) $6\frac{8}{18} + 9\frac{1}{6} =$

28) $30\frac{24}{26} + 4\frac{2}{78} =$

29) $1\frac{2}{15} \div 17 =$

30) $14\frac{2}{15} + 2\frac{3}{45} =$

31) $8\frac{3}{14} \times 1\frac{1}{2} =$

32) $4\frac{26}{30} \div 146 =$

33) $1\frac{2}{3} - \frac{7}{8} =$

34) $18 - 9\frac{11}{23} =$

35) $43\frac{2}{9} - 12\frac{1}{8} =$

36) $1\frac{3}{10} + 9\frac{16}{25} =$

37) $3\frac{6}{7} + 9\frac{1}{6} =$

38) $1\frac{1}{2} \div 2\frac{3}{7} =$

39) $28\frac{7}{81} - \frac{5}{9} =$

40) $12 + 9\frac{12}{13} =$

41) $\frac{2}{11} \times 1\frac{1}{10} =$

42) $\frac{3}{19} \times \frac{38}{39} =$

43) $\frac{2}{13} + \frac{4}{10} =$

44) $8\frac{2}{7} \div \frac{2}{21} =$

45) $\frac{33}{44} \times \frac{66}{77} =$

46) $3\frac{2}{5} \div 3\frac{4}{10} =$

47) $71 - 1\frac{4}{9} =$

48) $34\frac{12}{35} + 18 =$

REVIEW OF BASIC OPERATIONS

1) $243 + 2,003 =$

2) $4.5 \times .002 =$

3) $\frac{1}{9} + \frac{3}{7} =$

4) $3 - 0.453 =$

5) $13,000 - 279 =$

6) $52 + 0.52 =$

7) $2.09 \times 5.9 =$

8) $\frac{3}{10} \times \frac{5}{6} =$

9) $203 + 1,002 + 32 =$

10) $3.001 \times 2.03 =$

11) $2.6 + 12.92 + 5.6 =$

12) $0.208 - 0.091 =$

13) $0.03 + 0.112 + 1.1 =$

14) $3\frac{3}{11} \div 2\frac{1}{5} =$

15) $7 \times 1\frac{5}{6} =$

16) $0.002 - 0.0019 =$

17) $2.3 - 0.0342 =$

18) $5,202 \div 9 =$

19) $4,100 \times 200 =$

20) $\frac{7}{24} \div \frac{14}{36} =$

21) $19 - 7\frac{10}{17} =$

22) $1\frac{5}{8} + 2\frac{1}{6} =$

23) $0.08 + 1.3 =$

24) $14\frac{1}{3} + 13\frac{4}{27} =$

25) $1\frac{7}{9} \times 1\frac{7}{8} =$

26) $83 + 211 + 934 =$

27) $\frac{7}{11} - \frac{1}{22} =$

28) $4\frac{2}{3} \times 3\frac{2}{3} =$

29) $20,031 - 19,823 =$

30) $787,800 \div 78 =$

31) $19\frac{4}{11} + 2 =$

32) $50,010 - 2,992 =$

33) $\frac{2}{3} \times \frac{12}{16} =$

34) $7 + 5.6 + 0.09 =$

35) $1.875 \div 0.15 =$

36) $0.12 \times 0.005 =$

37) $1,890 \div 24 =$

38) $33\frac{2}{11} - \frac{9}{22} =$

39) $21,210 \div 20 =$

40) $518.4 \div 4.8 =$

41) $20 - 19\frac{2}{3} =$

42) $60,491 \div 60 =$

43) $934.33 + 2.6 =$

44) $99.6 - 4.06 =$

45) $6.061 \div 0.11 =$

46) $5\frac{7}{10} + 3\frac{9}{50} =$

47) $921 \times 0.44 =$

48) $1.001 \times 0.002 =$

Name _____

Class _____ Date _____

TEST FOR CHAPTER 1, Form A

1) Chris works as an electrician with the Shocksville Electric Company. He earns $6.25 per hour with time and a half after 40 hours. If he worked 43 hours Monday through Friday, how much would he earn? _____

2) One week Chris worked from 8:00 A.M. to 1:00 P.M., lunched until 1:30, and then worked until 6:00 P.M. How many hours did he work? _____

3) Mai receives 57¢ for every radio she wires at the Brandville Radio Company. How much will she earn if she completes the wiring on 105 radios? _____

▶ Round each amount to the nearer cent.

4) $19.623 = _____   5) $129.699 = _____   6) $5.607 = _____

▶ Change these percents to decimals.

7) 5% = _____   8) 17% = _____   9) 110% = _____

10) Mr. Abrams receives a 4% commission plus a salary of $75.00 per week. If his total sales for one week were $2880.00, how much did he earn? _____

11) Martinez, a car salesman, sold a car for $5400. How much did he earn with a 1.5% commission? _____

▶ Compute the take-home pay for these two employees.

	Gross Pay	Federal Tax	State Tax	Soc. Sec. Tax	Retirement	Health Ins.	Credit Union	Take-Home Pay
12)	$773	$115	$32	$47.38	$25	$9.25	0	_____
13)	$410	$95.75	$17.35	$20	$18.25	0	$7.89	_____

Name _____

Class _____ Date _____

TEST FOR CHAPTER 1, Form B

1) Wanda works as an electrician with the Current Electric Company. She earns $7.25 per hour with time and a half after 40 hours. If she worked 44 hours Monday through Friday, how much would she earn?   _____

2) One week Wanda worked from 8:00 A.M. to 1:00 P.M., lunched until 1:30, and then worked until 5:30 P.M. How many hours did she work?   _____

3) Pam receives 49¢ for every radio she wires at the Benton Radio Company. How much will she earn if she completes the wiring on 128 radios?   _____

▶ Round each amount to the nearer cent.

4) $18.937 = _____   5) $138.026 = _____   6) $2.509 = _____

▶ Change these percents to decimals.

7) 6% = _____   8) 19% = _____   9) ½% = _____

10) Mr. Garcia receives a 5% commission plus a salary of $85.00 per week. If his total sales for one week were $3912.00, how much did he earn?   _____

11) Bruce, a car salesman, sold a car for $6400. How much did he earn with a 1.4% commission?   _____

▶ Compute the take-home pay for these two employees.

	Gross Pay	Federal Tax	State Tax	Soc. Sec. Tax	Retire-ment	Health Ins.	Credit Union	Take-Home Pay
12)	$712	$34	$42.37	$43.37	$32	$8.89	0	_____
13)	$412	$93.75	$21	$17.45	$19	$6.10	$8.95	_____

Name _____

Class _____ Date _____

TEST FOR CHAPTER 2, Form A

▶ Write these as cents instead of dollars.

1) $4.65 = _____   2) $0.72 = _____   3) $0.12 = _____

▶ Write these as dollars instead of cents.

4) 53¢ = _____   5) 195¢ = _____   6) 3¢ = _____

Colossal Food Center

Potatoes	10 lb.	**$1.79**	**Milk** .... gal.	**$2.29**
Peaches, sliced	12-oz. can	**3/100**	**Butter** .... lb.	**2⁹⁸**
Corn	20-oz. pkg.	**2/89**	**Eggs** .... doz.	**$1.00**
Carrots	12-oz. pkg.	**49¢**	**Apple Pie** .... 37-oz. pkg.	**$1.35**
Lake's Salt	26-oz. box	**20¢**	**Bread** .... 20-oz. loaf	**89**
Smith's Popcorn	16-oz. can	**28¢**	**Shiny Car Wash** ...... 11-oz. can	**1⁵⁵**
Topper Popcorn	20-oz. pkg.	**42¢**	**Reddy Car Wash** ...... 10-oz. btl.	**1.30**

7) How much is one package of corn? _____

8) Mrs. Valle bought ten pounds of potatoes, two apple pies, and one can of peaches. How much was her total bill? _____

9) If Mrs. Valle gave the salesclerk twenty dollars, how much was her change? _____

10) Find the unit price of Smith's popcorn. How much does it cost per ounce.? _____

11) Find the unit price of Topper popcorn. How much does it cost per ounce? _____

12) Oscar used a 20¢-off coupon when he bought Reddy car wash. How much did the car wash cost per ounce.? _____

13) Mabel used a 20¢-off coupon when she bought Shiny car wash. How much did the car wash cost per ounce? _____

Name _____

Class _____ Date _____

TEST FOR CHAPTER 2, Form B

▶ Write these as cents instead of dollars.

1) $3.45 = _____   2) $0.45 = _____   3) $0.03 = _____

▶ Write these as dollars instead of cents.

4) 46¢ = _____   5) 193¢ = _____   6) 4¢ = _____

Colossal Food Center

Potatoes	10 lb.	**$1.58**	**Milk**	gal.	**222**
Peaches	12-oz. can	**2/45**	**Butter**	lb.	**2⁹⁹**
Corn	20-oz. pkg.	**3/100**	**Eggs**	doz.	**$1.29**
Carrots	12-oz. pkg.	**59¢**	**Apple Pie**	37-oz. pkg.	**$1.32**
Lake's Salt	26-oz. box	**30¢**	**Bread**	20-oz. loaf	**79**
Smith's Popcorn	16-oz. can	**26¢**	**Shiny Car Wash**	11-oz. can	**1⁵⁴**
Topper Popcorn	20-oz. pkg.	**34¢**	**Reddy Car Wash**	10-oz. btl.	**1.29**

---

7) How much is one package of corn? _____

8) Mr. Davis bought ten pounds of potatoes, two apple pies, and one can of peaches. How much was his total bill? _____

9) If Mr. Davis gave the salesclerk twenty dollars, how much was his change? _____

10) Find the unit price of Smith's popcorn. How much does it cost per ounce? _____

11) Find the unit price of Topper popcorn. How much does it cost per ounce? _____

12) Kathy used a 30¢-off coupon when she bought Reddy car wash. How much did the car wash cost per ounce? _____

13) Bill used a 30¢-off coupon when he bought Shiny car wash. How much did the car wash cost per ounce? _____

Name _____

Class _____ Date _____

TEST FOR CHAPTER 3, Form A

1) What is the total cost of a pair of shoes priced at $19.82, with a 5% sales tax? _____

2) What is the total cost of 3 pairs of socks priced at $1.29 a pair, with a 6% sales tax? _____

3) A stereo regularly costs $75.95. Bob bought it on sale for $68.35. How much did he save? _____

4) A jogging outfit normally sells for $45.95. How much will the outfit cost after a 20% discount? _____

5) A shirt that normally sells for $23.00 before discount was priced at $18.00 after discount. What is the rate of the discount? _____

▶ Use this table for problems 6 and 7.

Shipping Weight and Charges	
Up to 1 pound $3.04	6 lbs., 1 oz. to 12 lbs $8.99
1 lb., 1 oz. to 2 lbs.............3.56	Over 12 pounds 13.87
2 lbs., 1 oz. to 6 lbs............5.59	

6) A catalog order weighs ten pounds. What is the shipping charge? _____

7) A catalog order weighs one pound. What is the shipping charge? _____

8) If a fabric costs $5.60 a yard, how much will $7\frac{4}{5}$ yards cost? _____

9) The new balance of a charge account is $135.00. The minimum payment is $20.00. If the interest rate is 1.5%, how much will the new balance be next month? _____

10) Marion placed a coat on layaway. The coat cost $125. Marion made a 20% deposit. How much will the balance be? _____

Name _____

Class _____ Date _____

TEST FOR CHAPTER 3, Form B

1) What is the total cost of a pair of shoes priced at $18.95, with a 6% sales tax? _____

2) What is the total cost of 3 pairs of socks priced at $1.35 a pair, with a 7% sales tax? _____

3) A stereo regularly costs $89.90. Jim bought it on sale for $65.00. How much did he save? _____

4) A jogging outfit normally sells for $35.85. How much will the outfit cost after a 15% discount? _____

5) A shirt that normally sells for $21.00 before discount was priced at $17.00 after discount. What is the rate of the discount? _____

▶ Use this table for problems 6 and 7.

Shipping Weight and Charges	
Up to 1 pound $3.04	6 lbs., 1 oz. to 12 lbs $7.85
1 lb., 1 oz. to 2 lbs..............3.46	Over 12 pounds $13.77
2 lbs., 1 oz. to 6 lbs............5.49	

6) A catalog order weighs eleven pounds. What is the shipping charge? _____

7) A catalog order weighs two pounds. What is the shipping charge? _____

8) If a fabric costs $5.75 a yard, how much will $7\frac{4}{5}$ yards cost? _____

9) The new balance of a charge account is $145.00. The minimum payment is $30.00. If the interest rate is 1.5%, how much will the new balance be next month? _____

10) Maria placed a coat on layaway. The coat cost $125. Maria made a 10% deposit. How much will the balance be? _____

Name _____

Class _____ Date _____

METERS TO READ

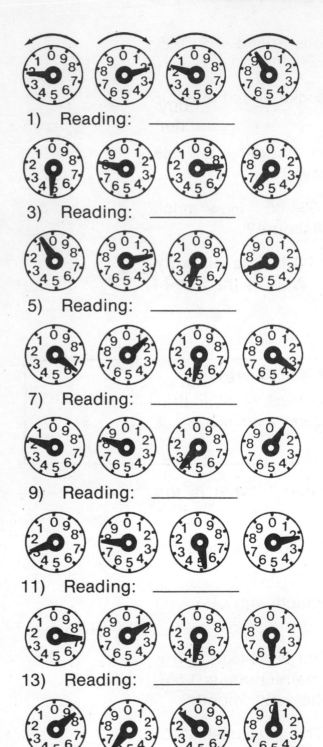

1) Reading: _____

3) Reading: _____

5) Reading: _____

7) Reading: _____

9) Reading: _____

11) Reading: _____

13) Reading: _____

15) Reading: _____

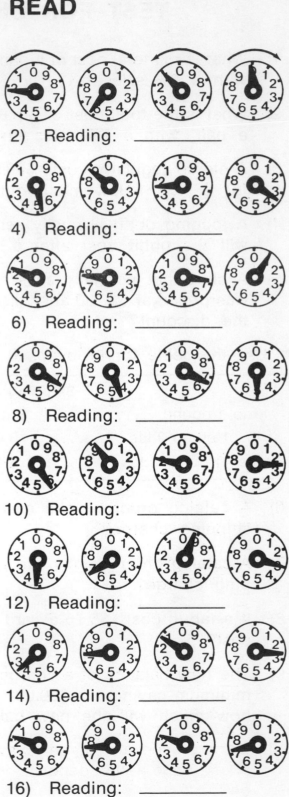

2) Reading: _____

4) Reading: _____

6) Reading: _____

8) Reading: _____

10) Reading: _____

12) Reading: _____

14) Reading: _____

16) Reading: _____

Name _____

Class _____ Date _____

TEST FOR CHAPTER 4, Form A

1) Eric's income is $566 every two weeks. Use the Renter's rule to compute the most Eric should spend for rent each month. _____

2) Ms. Burke earns $12,000 per year and is planning to buy a house. Use the Banker's rule to determine her maximum allowable mortgage. _____

3) Leo wants to buy a house for $48,000 with a 20% down payment. How much will his down payment be? _____

How much must he finance? _____

4)    5)

   March 31 reading _____   April 30 reading _____

6) Compare the two electric meter readings above. How many units of electricity were used during the month of April? _____

7) The monthly flat rate for Tim's telephone was $14.95. He was charged $1.85 for a long distance call and $1.76 for taxes. What was the total bill? _____

8) Find Jessica's utility expenses. The following amounts are due:

   Telephone, $18.56   Water, $2.73
   Gas, $47.84   Electricity, $39.92   _____

Area	Construction	Coverage Rate
A	Brick	.37%
	Wood Frame	.45%
B	Brick	.41%

▶ Use the table above to find the annual homeowners payment for:

9) A wood frame house in Area A whose value is $56,800. _____

10) A brick house in Area B whose value is $68,000. _____

Name _____

Class _____ Date _____

TEST FOR CHAPTER 4, Form B

1) Amy's income is $588 every two weeks. Use the Renter's rule to compute the most Amy should spend for rent each month.   _____

2) Mr. Morris earns $15,000 per year and is planning to buy a house. Use the Banker's rule to determine his maximum allowable mortgage.   _____

3) Joyce wants to buy a house for $49,500 with a 20% down payment. How much will her down payment be?   _____

   How much must she finance?   _____

4)

   March 31 reading _____

5)

   April 30 reading _____

6) Compare the two electric meter readings above. How many units of electricity were used during the month of April?   _____

7) The monthly flat rate for Jason's telephone was $13.75. He was charged $1.75 for a long distance call and $1.82 for taxes. What was his total bill?   _____

8) Find Rebecca's utility expenses. The following amounts are due:

   Telephone, $19.75   Water, $2.52
   Gas, $34.58   Electricity, $47.35   _____

Area	Construction	Coverage Rate
A	Brick	.36%
	Wood Frame	.46%
B	Brick	.42%

▶ Use the table above to find the annual homeowners payment for:

9) A wood frame house in Area A whose value is $55,900.   _____

10) A brick house in Area B whose value is $72,000.   _____

TEST FOR CHAPTER 5, Form A

1) Donna bought a new car whose base price was $5,290.00. She ordered a radio for $75, an automatic transmission for $450, and paid $507 for transportation and handling. What was the total price?   _____

2) Determine the actual cost, including interest, of a $2,450 car with a down payment of $200 and 48 monthly payments of $75 each.   _____

3)

Liability Insurance Rates

Area	Personal Injury			Property Damage			
	10/20	20/40	40/80	5	10	25	50
High Risk	110	154	173	72	84	99	148
Average Risk	67	71	89	56	69	77	98

Joan wants to buy a 20/40/10 liability insurance policy. How much will the total premium be if she lives in an average risk area?   _____

4) Norman drove his car on a vacation trip. His odometer read 1270.3 when he began and 1740.3 at the end. If he used 25 gallons of gas, how many miles per gallon did he get?   _____

5) If gas costs $0.98 a gallon, what is the cost of filling a 25-gallon tank?   _____

6) What is the range of Dan's car? Its highway EPA rating is 40 mpg, and the tank holds 12 gallons.   _____

7) Melissa drove 522.5 miles in 9 hours and 30 minutes. What was her average speed?   _____

8) Kirby made a 416-mile trip with an average speed of 52 miles per hour. What was his travel time?   _____

9) How many gallons of gas will Carlos need to drive 483 miles? His car has an EPA rating of 21 miles per gallon.   _____

10) Find Julio's repair expenses. His car needs a new muffler that costs $42.75. It will take two hours of labor at $25 per hour. There will be a 5% sales tax on the muffler.   _____

Name _____

Class _____ Date _____

TEST FOR CHAPTER 5, Form B

1) Amanda bought a new car whose base price was $5,380.00. She ordered a radio for $85, an automatic transmission for $475, and paid $422 for transportation and handling. What was the total price? _____

2) Determine the actual cost, including interest, of a $2,670 car with a down payment of $300 and 48 monthly payments of $73 each. _____

3)

Liability Insurance Rates

Area	Personal Injury			Property Damage			
	10/20	20/40	40/80	5	10	25	50
High Risk	110	154	173	72	84	99	148
Average Risk	67	71	89	56	69	77	98

Michelle wants to buy a 20/40/25 liability insurance policy. How much will the total premium be if she lives in an average risk area? _____

4) Juan drove his car on a vacation trip. His odometer read 1460.3 when he began and 1830.2 at the end. If he used 20 gallons of gas, how many miles per gallon did he get? _____

5) If gas costs $1.35 a gallon, what is the cost of filling a 20-gallon tank? _____

6) What is the range of Kelly's car? Its highway EPA rating is 45 mpg, and the tank holds 15 gallons. _____

7) Heather drove 498.5 miles in 10 hours and 30 minutes. What was her average speed? _____

8) Joshua made a 486-mile trip with an average speed of 54 miles per hour. What was his travel time? _____

9) How many gallons of gas will Brian need to drive 513 miles? His car has an EPA rating of 22 miles per gallon. _____

10) Find Stephen's repair expenses. His car needs a new muffler that costs $43.77. It will take two hours of labor at $23 per hour. There will be a 5% sales tax on the muffler. _____

Name _____

Class _____ Date _____

TEST FOR CHAPTER 6, Form A

1) For breakfast Kevin had a cup of milk, 2 medium eggs, 2 slices of toast, and 1 cup of orange juice. Use the Calorie Chart to compute the total calories in Kevin's meal.   _____

Calorie Chart

Milk, 1 cup . . . . 150 calories	Toast, slice . . . . 65 calories
Egg, medium . . . . 95 calories	Orange juice, cup . . . . 120 calories

2) Errol bought an apple pie cut into fifths.   He ate the whole pie. If each piece contained 372 calories, how many calories did he consume?   _____

3) If a 12-ounce glass of ginger ale contains 115 calories, how many calories are in 10 ounces?   _____

4) If 8 ounces of tomato soup contain 90 calories, how many calories are contained in 12 ounces?   _____

5) Eight ounces of orange juice contains 0.02 mg of iron. If the RDA of iron is 18 mg, what percent of the RDA of iron is provided by the orange juice?   _____

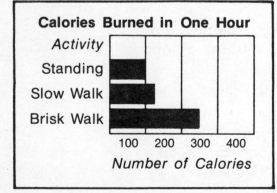

6) How many calories are burned in a brisk, 2-hour walk?   _____

7) If a pound of body fat contains about 3500 calories, how long will it take to lose one pound by standing?   _____

8) A recipe that serves six calls for 5 ounces of mushroom sauce. Compute the amount of sauce needed to serve ten people.   _____

9) What time must a roast be placed in the oven to be ready by 5:30 if the roast takes 2 hours and 10 minutes for cooking?   _____

10) What time must these items be placed in the oven to be ready by 5:45?

The roast needs 3 hours and 20 minutes.   _____

The vegetables need 25 minutes.   _____

Name _____

Class _____ Date _____

TEST FOR CHAPTER 6, Form B

1) For breakfast Katie had 2 cups of milk, 3 medium eggs, 3 slices of toast, and 1 cup of orange juice. Use the Calorie Chart to compute the total calories in Katie's meal.   _____

Calorie Chart

Milk, 1 cup . . . . 150 calories	Toast, slice . . . . 65 calories
Egg, medium . . . . 95 calories	Orange juice, cup . . . . 120 calories

2) Gene bought an apple pie cut into fifths. He ate the whole pie. If each piece contained 123 calories, how many calories did he consume?   _____

3) If a 16-ounce glass of ginger ale contains 195 calories, how many calories are in 10 ounces?   _____

4) If 8 ounces of celery soup contain 72 calories, how many calories are contained in 12 ounces?   _____

5) Eight ounces of orange juice contains 1 mg of niacin. If the RDA of niacin is 20 mg, what percent of the RDA of niacin is provided by the orange juice?   _____

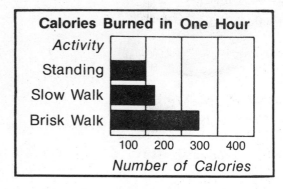

Calories Burned in One Hour

Activity
Standing
Slow Walk
Brisk Walk

100 200 300 400
Number of Calories

6) How many calories are burned in a brisk, 3-hour walk?   _____

7) If a pound of body fat contains about 3500 calories, how long will it take to lose one pound by walking slowly?   _____

8) A recipe that serves six calls for 5 ounces of mushroom sauce. Compute the amount of sauce needed to serve 8 people.   _____

9) What time must a roast be placed in the oven to be ready by 4:45 if the roast takes 3 hours and 10 minutes for cooking?   _____

10) What time must these items be placed in the oven to be ready by 5:45?
The roast needs 2 hours and 10 minutes.   _____

The vegetables need 35 minutes.   _____

TEST FOR CHAPTER 7, Form A

1) Sophie bought a couch for $364.00 after a 20% discount. What was the original price? _____

2) Mr. Pepper bought a desk on sale for $209.60. If the original cost was $262.00, what was the percent of discount? _____

3) Barbara bought a kitchen table on January 24, using the 90-day plan. By what day must the total be paid? It was not a leap year. _____

4) Compute the perimeter and area of this figure.

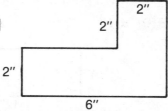

Perimeter _____

Area _____

5) How many quarts of paint must Tim buy to paint the walls of a room with a length of 9 feet, a width of 12 feet, and a height of 8 feet? The paint will cover 100 square feet per quart. _____

6) How many square feet of wallpaper will Linda need to paper a room measuring 10′ × 10′ with a height of 8 feet? _____

7) How much will it cost to cover a floor that measures 12 feet by 13 feet with 12″ × 12″ tiles costing $1.55 each? _____

8) Compute the amount of molding needed for a room that measures $12 \times 13\frac{1}{2}$ feet. _____

9) The Hendersons are planning to add a family room to their existing house. The family room will cost $52 per square foot. Compute the cost of the new room if the room measures 13 by 15 feet. _____

10) Compute the length of fence needed to enclose this yard.

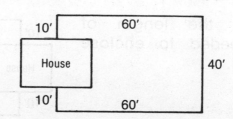

_____

Name _____

Class _____ Date _____

TEST FOR CHAPTER 7, Form B

1) Debbie bought a couch for $395.00 after a 20% discount. What was the original price?   _____

2) Miss Jennie bought a desk on sale for $178.16. If the original cost was $262.00, what was the percent of discount?   _____

3) Linus bought a kitchen table on January 22, using the 90-day plan. By what day must the total be paid? It was not a leap year.   _____

4) Compute the perimeter and area of this figure.

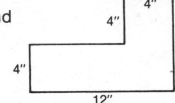

Perimeter _____

Area _____

5) How many quarts of paint must Ann buy to paint the walls of a room with a length of 9 feet, a width of 13 feet, and a height of 8 feet? The paint will cover 100 square feet per quart.   _____

6) How many square feet of wallpaper will Patty Ann need to paper a room measuring 10' × 11' with a height of 8 feet?   _____

7) How much will it cost to cover a floor that measures 13 feet by 13 feet with 12" × 12" tiles costing $1.55 each?   _____

8) Compute the amount of molding needed for a room that measures $14 \times 13\frac{1}{2}$ feet.   _____

9) The Jacksons are planning to add a family room to their existing house. The family room will cost $49 per square foot. Compute the cost of the new room if the room measures 14 by 15 feet.   _____

10) Compute the length of fence needed to enclose this yard.   _____

Name _____

Class _____ Date _____

TEST FOR CHAPTER 8, Form A

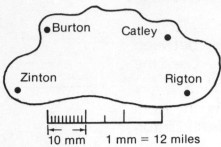

Burton Catley

Zinton Rigton

10 mm   1 mm = 12 miles

1) What is the distance between Rigton and Zinton? Use the scale. _____

2) How much farther is it from Zinton to Rigton than from Zinton to Burton? _____

3) Tom is leaving by train at 10:50 A.M. The train is to arrive in Markton at 2:45 P.M. How long will the trip take? _____

4) Compute the hotel expenses of a family of four on vacation for five days. Include a 15% tip for the excellent service. _____

   Room Rates: Single, $39.50; Double, $44.50; Suite, $87.50

5) How many dollars can be exchanged for 131 West German marks? _____

   2.58 marks = 1 U.S. dollar
   1 mark = $0.39 in U.S. dollars

6) When Jacob visited the United States, he saw a radio that he liked. It was priced at $78. How much would it cost in marks?

7) Malinda wants to rent a compact car for 8 days. Hurts rents these cars for $48 per day and $340 per week. Compute the cost of each plan for her. Which plan is more economical? _____

8) Mike parked his tricycle at the Tiny Tots Parking Lot from 10:15 A.M. to 3:10 P.M. It cost $1.50 for the first hour and $1.00 for each additional hour or partial hour. How much did he have to pay? _____

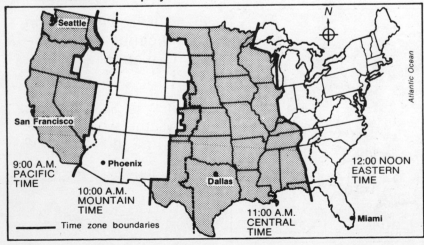

9) If it is 3:40 P.M. in Phoenix, what time is it in Miami? _____

10) If it is 2:20 A.M. in Miami, what time is it in San Francisco?

Name _____

Class _____ Date _____

TEST FOR CHAPTER 8, Form B

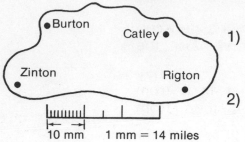

1) What is the distance between Rigton and Zinton? Use the scale. _____

2) How much farther is it from Zinton to Rigton than from Zinton to Burton? _____

3) Ryan is leaving by train at 10:50 A.M. The train is to arrive in Markton at 3:30 P.M. How long will the trip take? _____

4) Compute the hotel expenses of a family of four on vacation for four days. Include a 15% tip for the excellent service.

Room Rates: Single, $39.50; Double, $44.50; Suite, $87.50

5) How many dollars can be exchanged for 120 West German marks? _____

2.58 marks = 1 U.S. dollar
1 mark = $0.39 in U.S. dollars

6) When Karl visited the United States, he saw a radio that he liked. It was priced at $117. How much would it cost in marks? _____

7) Annette wants to rent a compact car for 9 days. Hurts rents these cars for $48 per day and $340 per week. Compute the cost of each plan for her. Which plan is more economical? _____ _____

8) Sean parked his tricycle at the Tiny Tots Parking Lot from 10:15 A.M. to 2:10 P.M. It costs $1.50 for the first hour and $1.00 for each additional hour or partial hour. How much did he have to pay? _____

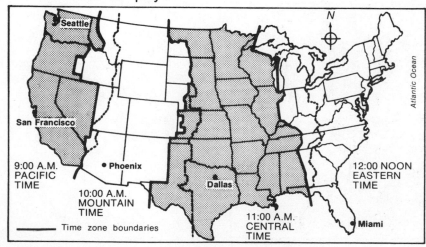

9) If it is 2:40 P.M. in Phoenix, what time is it in Miami? _____

10) If it is 3:20 A.M. in Miami, what time is it in San Francisco? _____

Name _____

Class _____ Date _____

TEST FOR CHAPTER 9, Form A

1) Find Mike's average monthly income, based on these weekly incomes: $142, $146, $209, $175, $235.   _____

2) Find Cecil's average monthly income if his take-home pay is $382.00 every two weeks.   _____

3) Nancy Danver budgets her monthly income of $725.00 as shown in the chart below. Find the amount of money allotted for each item.

Housing, 27%	Transportation, 13%	Insurance, 5%	Gifts, 7%
Food, 20%	Entertainment, 5%	Health, 3%	
Clothes, 9%	Miscellaneous, 3%	Savings, 8%	

Housing = _____   4) Entertainment = _____

Food = _____   Miscellaneous = _____

Clothes = _____   Insurance = _____

Gifts = _____   Health = _____

Transportation = _____   Savings = _____

5) Brian's monthly net income is $435.70. He budgets 21% for food. Find the amount of money he budgets for food.   _____

6) Neal's family spends $1342 annually for food. Approximately what percent of his $14,910 annual net income is this?   _____

7) Jennifer earns $175 per week and spends her money on the following. Find the percent spent in each category. Round to nearest whole percent.

Records and tapes, $12.00 = ___%	Entertainment,	$25.00 = ___%
Savings, $50.00 = ___%	Hair and make-up,	$30.00 = ___%

8) At a sale Kim overspent her clothing allotment by $12. She allots 9% of her monthly income of $725 for clothes. How much will be available the next month for clothes?   _____

9) A circle has 360°. If the following budget were shown by a circle graph, how many degrees would each category take?

Food, 25% = _____ degrees

Housing, 15% = _____ degrees

Clothing, 20% = _____ degrees

Car, 10% = _____ degrees

Health, 10% = _____ degrees

Miscellaneous, 20% = _____ degrees

10) Make a circle graph for the budget in problem 9.

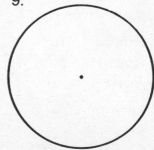

Name _____

Class _____ Date _____

TEST FOR CHAPTER 9, Form B

1) Find Sara's average monthly income, based on these weekly incomes: $135, $146, $209, $175. _____

2) Find James's average monthly income if his take-home pay is $427.00 every two weeks. _____

3) Marie budgets her monthly income of $642.00 as shown in the chart below. Find the amount of money allotted for each item.

Housing, 27%	Transportation, 13%	Insurance, 5%	Gifts, 7%
Food, 20%	Entertainment, 5%	Health, 3%	
Clothes, 9%	Miscellaneous, 3%	Savings, 8%	

Housing = _____   4) Entertainment = _____

Food = _____   Miscellaneous = _____

Clothes = _____   Insurance = _____

Gifts = _____   Health = _____

Transportation = _____   Savings = _____

5) Jeff's monthly income is $535.70. He budgets 21% for food. Find the amount of money he budgets for food. _____

6) Cal's family spends $1250 annually for food. Approximately what percent of his $14,910 annual net income is this? _____

7) Amanda earns $185 per week and spends her money on the following. Find the percent spent in each category. Round to the nearest whole percent.

Records and tapes,	$12.00 = ___%	Entertainment,	$25.00 = ___%
Savings,	$50.00 = ___%	Hair and make-up,	$30.00 = ___%

8) At a sale Erin overspent her clothing allotment by $10. She allots 9% of her monthly income of $725 for clothes. How much will be available the next month for clothes? _____

9) A circle has 360°. If the following budget were shown by a circle graph, how many degrees would each category take?

Food, 25% = _____ degrees

Housing, 15% = _____ degrees

Clothing, 20% = _____ degrees

Car, 5% = _____ degrees

Health, 15% = _____ degrees

Miscellaneous, 20% = _____ degrees

10) Make a circle graph for the budget in problem 9.

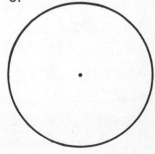

Name _____

Class _____ Date _____

CHECKS

	Dollars	Cents
No. _____		
_____ 19___		
To _____		
BAL. FWD.		
DEPOSITS		
TOTAL		
THIS CHECK		
BALANCE		
DEDUCTIONS		
BAL. FWD.		

NO. _____

7-89 / 520

_____ 19___

PAY TO THE
ORDER OF _____ $ _____

_____ DOLLARS

RIVER BANK OF COLUMBUS

FOR _____   _____

⑆052000896⑆0772‖752 2410 2‖⑈

	Dollars	Cents
No. _____		
_____ 19___		
To _____		
BAL. FWD.		
DEPOSITS		
TOTAL		
THIS CHECK		
BALANCE		
DEDUCTIONS		
BAL. FWD.		

NO. _____

7-89 / 520

_____ 19___

PAY TO THE
ORDER OF _____ $ _____

_____ DOLLARS

RIVER BANK OF COLUMBUS

FOR _____   _____

⑆052000896⑆0772‖752 2410 2‖⑈

	Dollars	Cents
No. _____		
_____ 19___		
To _____		
BAL. FWD.		
DEPOSITS		
TOTAL		
THIS CHECK		
BALANCE		
DEDUCTIONS		
BAL. FWD.		

NO. _____

7-89 / 520

_____ 19___

PAY TO THE
ORDER OF _____ $ _____

_____ DOLLARS

RIVER BANK OF COLUMBUS

FOR _____   _____

⑆052000896⑆0772‖752 2410 2‖⑈

Name _____

Class _____ Date _____

TRANSACTION REGISTER FORM

This record form can be used to keep a record of all check and non-check transactions.

									BAL. FWD.	
DATE	TRANS NUMBER	TYPE OF TRANS	DESCRIPTION		AMT OF TRANS(−)	AMT OF DEPOSIT(+)	FEE(−)	TAX ITEM		

CHECK = CK   DEPOSIT = D   DEBIT CARD = DC   HOME COMPUTER = HC   MACHINE = M   PHONE = PH

Double-line Method

Balance is recorded after each entry in the gray area of the balance column.

DATE	TRANS NUMBER	TYPE OF TRANS	DESCRIPTION	AMT OF TRANS(−)	AMT OF DEPOSIT(+)	FEE( )	TAX ITEM	BAL. FWD.	
								600	00
3/19	101	PH	Auto Insurance	100.00				−100	00
								500	00
3/19	102	CK	S. H. Kirk	55.80				− 55	80
			Mary's gift					444	20
3/22	103	M	Cash	60.00				− 60	00
								384	20
3/23		D	Deposit		150.00			+150	00
			from paycheck					534	20

Name _____

Class _____ Date _____

TEST FOR CHAPTER 10, Form A

1) Jeanne is making a loan of $500 to her roommate. The room-
mate will pay 7% simple interest per year. How much
interest will Jeanne earn in 8 months? _____

2) A new savings account of $750.00 earns 6% interest com-
pounded quarterly. Compute the balance at the end of one
year. _____

3) Find the approximate number of years it will take to double
your money if your beginning balance is $1,000.00 and earns
10% interest compounded quarterly. _____

4) Write in words these amounts as they would appear on a check.

a) $345.72 _____

b) $1,022.00 _____

5) Find the balance after each transaction. The beginning balance was $752.00.

No.	Date	Transaction	Description	Amount	Balance
31	May 2	Check	Telephone	$19.75	_____
32	May 3	Check	Gas & Elec.	$102.51	_____
	May 3	Deposit	Paycheck	$352.00	_____
33	May 5	Check	Rent	$250.00	_____
34	May 7	Check	Dentist	$78.00	_____

6) Reconcile the following checking account.

Bank balance, $1742.50   Checkbook balance, $1655.00

Unprocessed Deposits, $263.00   Does it balance? _____

Unreturned checks: $5.50, $122.00, $8.00,   If not, what is
$63.00, $9.00, $75.00, $19.50, $48.00   the difference? _____

7) Al bought stock at $17\frac{5}{8}$ and sold it for $21\frac{1}{4}$. What was his
profit? _____

8) Compute the profit or loss for these stocks.

a) Bought at $25\frac{3}{8}$. Sold at $18\frac{3}{4}$. _____ Profit or _____ Loss

b) Bought at $16\frac{1}{8}$. Sold at $20\frac{1}{4}$. _____ Profit or _____ Loss

9) Wilmer bought Wiggit stock at $15\frac{1}{8}$ and sold it for 18. What
was the percent of increase? _____

10) Marie owns 135 shares of stock. Her dividend is $0.50 per
share. How many shares can she buy of a different stock
selling for 11? _____

Name _____

Class _____ Date _____

TEST FOR CHAPTER 10, Form B

1) Nicole is making a loan of $400 to her roommate. The room-mate will pay 7% simple interest per year. How much interest will Nicole earn in 8 months? _____

2) A new savings account of $850.00 earns 6% interest compounded quarterly. Compute the balance at the end of one year. _____

3) Find the approximate number of years it will take to double your money if your beginning balance is $2,000.00 and earns 10% interest compounded quarterly. _____

4) Write in words these amounts as they would appear on a check.
   a) $235.72 _____
   b) $1,032.00 _____

5) Find the balance after each transaction. The beginning balance was $752.00.

No.	Date	Transaction	Description	Amount	Balance
31	May 2	Check	Telephone	$21.34	_____
32	May 3	Check	Gas & Elec.	$102.51	_____
	May 3	Deposit	Paycheck	$345.00	_____
33	May 5	Check	Rent	$250.00	_____
34	May 7	Check	Dentist	$78.00	_____

6) Reconcile the following checking account.
   Bank Balance, $1743.50   Checkbook Balance, $1655.00
   Unprocessed Deposits, $261.00   Does it balance? _____
   Unreturned checks: $5.50, $122.00,   If not, what is
   $8.00, $63.00, $9.00, $75.00, $19.50, $48.00   the difference? _____

7) Tim bought stock at $16\frac{5}{8}$ and sold it for $21\frac{1}{4}$. What was his profit? _____

8) Compute the profit or loss for these stocks.
   a) Bought at $20\frac{3}{8}$. Sold at $23\frac{3}{4}$. _____ Profit or _____ Loss
   b) Bought at $17\frac{1}{8}$. Sold at $12\frac{1}{4}$. _____ Profit or _____ Loss

9) Chaim bought Wiggit stock at $14\frac{1}{8}$ and sold it for 18. What was the percent of increase? _____

10) Wendy owns 135 shares of stock. Her dividend is $0.50 per share. How many shares can she buy of a different stock selling for 11? _____

FEDERAL INCOME TAX

Form 1040A

You can use Form 1040A for:

Any of four filing statuses

All exemptions you are entitled to

All qualified dependents

Income from:
  Wages, salaries, tips
  Interest and dividends
  Unemployment compensation
  Less than $50,000 in taxable income

Partial charitable contributions deduction

Deduction for a married couple when both work

Partial credit for political contributions

Earned income credit

You may get tax forms from many banks, post offices, public libraries, or from your local Internal Revenue Service Center.

▶ Fill out a 1040A form for each of these cases.

1) Scott and Bea Free of 143 Lovers Lane, Sweetville, Oklahoma, 74101, have two children, Bill and Spring. Together Bea and Scott earned $26,754.17 in salaries. Their savings account yielded interest of $257.10. Their stock paid dividends of $152.89. They gave $400 to qualified charities. Scott is a sheet-metal worker whose Social Security number is 123-45-6789. Bea, a registered electrician, has a Social Security number of 987-65-4321. A total of $4636.10 was withheld from their pay. They will file jointly.

2) John and Mary Dough have one child, Cookie, and will file a joint tax return. John is a teacher who earned $17,123.42. Mary is a social worker who earned $16,425.98. Their bank account yielded interest of $72.43, and they received stock dividends of $19.42. John's mother, Etta, lives with the Doughs at 1 Main Street, Washington, D.C. 20011, and is totally dependent on them since she has no income of her own. They gave $600 to qualified charities. A total of $6000 was withheld from John's (S.S.# 202-33-4611) and Mary's (S.S.# 211-43-8051) pay.

Name _____

Class _____ Date _____

TEST FOR CHAPTER 11, Form A

1) Write the numeral that means the same as "three hundred fifty-two billion, six hundred thirty-five million." _____

2) Write the number 265,401,000,000 in words.

_____

3) Use the tax table on page 230 of the textbook. Find the tax due for these two taxpayers:

Filing Status	Taxable Income	Tax Due
a) Single	$19,150	_____
b) Married, filing jointly	$19,225	_____

4) Ted's taxable income was $19,278. His employer withheld $3,715. Determine the amount to be refunded or to be paid. Ted is head of a household.

The refund is _____. OR the amount to be paid is _____.

5) Write these decimals in order from smallest to largest.

a) 1.23	1.52	1.203	_____	_____	_____
b) 10.06	1.07	11.1	_____	_____	_____
c) 5.32	5.308	5.4	_____	_____	_____
d) 7.02	.701	7.003	_____	_____	_____
e) .1	.101	1.1	_____	_____	_____

6) Find the assessed value of a home with a market value of $58,000 if the assessment rate is 42%. _____

7) Express the tax rate of $2.95 per $100 as a percent. _____

8) Pat and Don bought a home for $52,000. The property tax rate is $3.95 per $100. The assessment rate is 41%. How much is their property tax? _____

9) Find Pat and Don's effective tax rate. Use problem 8. _____

10) A $52,000 home has an assessment rate of 32% and a tax rate of $38 per $1,000. What is the effective tax rate? _____

What is the property tax? Use the effective tax method. _____

Name _____

Class _____ Date _____

TEST FOR CHAPTER 11, Form B

1) Write the numeral that means the same as "three hundred fifty-two billion, six hundred forty-two million." _____

2) Write the number 521,306,000,000 in words.

_____

3) Use the table on page 230 of the textbook. Find the tax due for these two taxpayers:

Filing Status	Taxable Income	Tax Due
a) Single	$20,100	_____
b) Married, filing jointly	$20,375	_____

4) Ted's taxable income was $20,905. His employer withheld $5,000. Determine the amount to be refunded or to be paid. Ted is head of a household.

The refund is _____.   OR   The amount to be paid is _____.

5) Write these decimals in order from smallest to largest.

a)	1.32	1.41	1.09	_____ _____ _____
b)	10.19	10.9	1.99	_____ _____ _____
c)	5.33	5.308	5.4	_____ _____ _____
d)	6.02	60.1	.600	_____ _____ _____
e)	.2	.202	2.2	_____ _____ _____

6) Find the assessed value of a home with a market value of $48,000 if the assessment rate is 43%   _____

7) Express the tax rate of $2.85 per $100 as a percent.   _____

8) Doug and Sue bought a home for $51,000. The property tax rate is $2.95 per $100. The assessment rate is 41%. How much is their property tax?   _____

9) Find Doug and Sue's effective tax rate. Use problem 8.   _____

10) A $51,000 home has an assessment rate of 36% and a tax rate of $37 per $1,000. What is the effective tax rate?   _____

What is the property tax? Use the effective tax method.   _____

Name _____

Class _____ Date _____

SALES TAX CHART

5% STATE SALES TAX

Amount of Sale	Tax	Amount of Sale	Tax	Amount of Sale	Tax	Amount of Sale	Tax	Amount of Sale	Tax	Amount of Sale	Tax
.20	.01										
.21 - .40	.02	10.01 - 10.20	.51	20.01 - 20.20	1.01	30.01 - 30.20	1.51	40.01 - 40.20	2.01	50.01 - 50.20	2.51
.41 - .60	.03	10.21 - 10.40	.52	20.21 - 20.40	1.02	30.21 - 30.40	1.52	40.21 - 40.40	2.02	50.21 - 50.40	2.52
.61 - .80	.04	10.41 - 10.60	.53	20.41 - 20.60	1.03	30.41 - 30.60	1.53	40.41 - 40.60	2.03	50.41 - 50.60	2.53
.81 - 1.00	.05	10.61 - 10.80	.54	20.61 - 20.80	1.04	30.61 - 30.80	1.54	40.61 - 40.80	2.04	50.61 - 50.80	2.54
Meals - 1.00	.05	10.81 - 11.00	.55	20.81 - 21.00	1.05	30.81 - 31.00	1.55	40.81 - 41.00	2.05	50.81 - 51.00	2.55
1.01 - 1.20	.06	11.01 - 11.20	.56	21.01 - 21.20	1.06	31.01 - 31.20	1.56	41.01 - 41.20	2.06	51.01 - 51.20	2.56
1.21 - 1.40	.07	11.21 - 11.40	.57	21.21 - 21.40	1.07	31.21 - 31.40	1.57	41.21 - 41.40	2.07	51.21 - 51.40	2.57
1.41 - 1.60	.08	11.41 - 11.60	.58	21.41 - 21.60	1.08	31.41 - 31.60	1.58	41.41 - 41.60	2.08	51.41 - 51.60	2.58
1.61 - 1.80	.09	11.61 - 11.80	.59	21.61 - 21.80	1.09	31.61 - 31.80	1.59	41.61 - 41.80	2.09	51.61 - 51.80	2.59
1.81 - 2.00	.10	11.81 - 12.00	.60	21.81 - 22.00	1.10	31.81 - 32.00	1.60	41.81 - 42.00	2.10	51.81 - 52.00	2.60
2.01 - 2.20	.11	12.01 - 12.20	.61	22.01 - 22.20	1.11	32.01 - 32.20	1.61	42.01 - 42.20	2.11	52.01 - 52.20	2.61
2.21 - 2.40	.12	12.21 - 12.40	.62	22.21 - 22.40	1.12	32.21 - 32.40	1.62	42.21 - 42.40	2.12	52.21 - 52.40	2.62
2.41 - 2.60	.13	12.41 - 12.60	.63	22.41 - 22.60	1.13	32.41 - 32.60	1.63	42.41 - 42.60	2.13	52.41 - 52.60	2.63
2.61 - 2.80	.14	12.61 - 12.80	.64	22.61 - 22.80	1.14	32.61 - 32.80	1.64	42.61 - 42.80	2.14	52.61 - 52.80	2.64
2.81 - 3.00	.15	12.81 - 13.00	.65	22.81 - 23.00	1.15	32.81 - 33.00	1.65	42.81 - 43.00	2.15	52.81 - 53.00	2.65
3.01 - 3.20	.16	13.01 - 13.20	.66	23.01 - 23.20	1.16	33.01 - 33.20	1.66	43.01 - 43.20	2.16	53.01 - 53.20	2.66
3.21 - 3.40	.17	13.21 - 13.40	.67	23.21 - 23.40	1.17	33.21 - 33.40	1.67	43.21 - 43.40	2.17	53.21 - 53.40	2.67
3.41 - 3.60	.18	13.41 - 13.60	.68	23.41 - 23.60	1.18	33.41 - 33.60	1.68	43.41 - 43.60	2.18	53.41 - 53.60	2.68
3.61 - 3.80	.19	13.61 - 13.80	.69	23.61 - 23.80	1.19	33.61 - 33.80	1.69	43.61 - 43.80	2.19	53.61 - 53.80	2.69
3.81 - 4.00	.20	13.81 - 14.00	.70	23.81 - 24.00	1.20	33.81 - 34.00	1.70	43.81 - 44.00	2.20	53.81 - 54.00	2.70
4.01 - 4.20	.21	14.01 - 14.20	.71	24.01 - 24.20	1.21	34.01 - 34.20	1.71	44.01 - 44.20	2.21	54.01 - 54.20	2.71
4.21 - 4.40	.22	14.21 - 14.40	.72	24.21 - 24.40	1.22	34.21 - 34.40	1.72	44.21 - 44.40	2.22	54.21 - 54.40	2.72
4.41 - 4.60	.23	14.41 - 14.60	.73	24.41 - 24.60	1.23	34.41 - 34.60	1.73	44.41 - 44.60	2.23	54.41 - 54.60	2.73
4.61 - 4.80	.24	14.61 - 14.80	.74	24.61 - 24.80	1.24	34.61 - 34.80	1.74	44.61 - 44.80	2.24	54.61 - 54.80	2.74
4.81 - 5.00	.25	14.81 - 15.00	.75	24.81 - 25.00	1.25	34.81 - 35.00	1.75	44.81 - 45.00	2.25	54.81 - 55.00	2.75
5.01 - 5.20	.26	15.01 - 15.20	.76	25.01 - 25.20	1.26	35.01 - 35.20	1.76	45.01 - 45.20	2.26	55.01 - 55.20	2.76
5.21 - 5.40	.27	15.21 - 15.40	.77	25.21 - 25.40	1.27	35.21 - 35.40	1.77	45.21 - 45.40	2.27	55.21 - 55.40	2.77
5.41 - 5.60	.28	15.41 - 15.60	.78	25.41 - 25.60	1.28	35.41 - 35.60	1.78	45.41 - 45.60	2.28	55.41 - 55.60	2.78
5.61 - 5.80	.29	15.61 - 15.80	.79	25.61 - 25.80	1.29	35.61 - 35.80	1.79	45.61 - 45.80	2.29	55.61 - 55.80	2.79
5.81 - 6.00	.30	15.81 - 16.00	.80	25.81 - 26.00	1.30	35.81 - 36.00	1.80	45.81 - 46.00	2.30	55.81 - 56.00	2.80
6.01 - 6.20	.31	16.01 - 16.20	.81	26.01 - 26.20	1.31	36.01 - 36.20	1.81	46.01 - 46.20	2.31	56.01 - 56.20	2.81
6.21 - 6.40	.32	16.21 - 16.40	.82	26.21 - 26.40	1.32	36.21 - 36.40	1.82	46.21 - 46.40	2.32	56.21 - 56.40	2.82
6.41 - 6.60	.33	16.41 - 16.60	.83	26.41 - 26.60	1.33	36.41 - 36.60	1.83	46.41 - 46.60	2.33	56.41 - 56.60	2.83
6.61 - 6.80	.34	16.61 - 16.80	.84	26.61 - 26.80	1.34	36.61 - 36.80	1.84	46.61 - 46.80	2.34	56.61 - 56.80	2.84
6.81 - 7.00	.35	16.81 - 17.00	.85	26.81 - 27.00	1.35	36.81 - 37.00	1.85	46.81 - 47.00	2.35	56.81 - 57.00	2.85
7.01 - 7.20	.36	17.01 - 17.20	.86	27.01 - 27.20	1.36	37.01 - 37.20	1.86	47.01 - 47.20	2.36	57.01 - 57.20	2.86
7.21 - 7.40	.37	17.21 - 17.40	.87	27.21 - 27.40	1.37	37.21 - 37.40	1.87	47.21 - 47.40	2.37	57.21 - 57.40	2.87
7.41 - 7.60	.38	17.41 - 17.60	.88	27.41 - 27.60	1.38	37.41 - 37.60	1.88	47.41 - 47.60	2.38	57.41 - 57.60	2.88
7.61 - 7.80	.39	17.61 - 17.80	.89	27.61 - 27.80	1.39	37.61 - 37.80	1.89	47.61 - 47.80	2.39	57.61 - 57.80	2.89
7.81 - 8.00	.40	17.81 - 18.00	.90	27.81 - 28.00	1.40	37.81 - 38.00	1.90	47.81 - 48.00	2.40	57.81 - 58.00	2.90
8.01 - 8.20	.41	18.01 - 18.20	.91	28.01 - 28.20	1.41	38.01 - 38.20	1.91	48.01 - 48.20	2.41	58.01 - 58.20	2.91
8.21 - 8.40	.42	18.21 - 18.40	.92	28.21 - 28.40	1.42	38.21 - 38.40	1.92	48.21 - 48.40	2.42	58.21 - 58.40	2.92
8.41 - 8.60	.43	18.41 - 18.60	.93	28.41 - 28.60	1.43	38.41 - 38.60	1.93	48.41 - 48.60	2.43	58.41 - 58.60	2.93
8.61 - 8.80	.44	18.61 - 18.80	.94	28.61 - 28.80	1.44	38.61 - 38.80	1.94	48.61 - 48.80	2.44	58.61 - 58.80	2.94
8.81 - 9.00	.45	18.81 - 19.00	.95	28.81 - 29.00	1.45	38.81 - 39.00	1.95	48.81 - 49.00	2.45	58.81 - 59.00	2.95
9.01 - 9.20	.46	19.01 - 19.20	.96	29.01 - 29.20	1.46	39.01 - 39.20	1.96	49.01 - 49.20	2.46	59.01 - 59.20	2.96
9.21 - 9.40	.47	19.21 - 19.40	.97	29.21 - 29.40	1.47	39.21 - 39.40	1.97	49.21 - 49.40	2.47	59.21 - 59.40	2.97
9.41 - 9.60	.48	19.41 - 19.60	.98	29.41 - 29.60	1.48	39.41 - 39.60	1.98	49.41 - 49.60	2.48	59.41 - 59.60	2.98
9.61 - 9.80	.49	19.61 - 19.80	.99	29.61 - 29.80	1.49	39.61 - 39.80	1.99	49.61 - 49.80	2.49	59.61 - 59.80	2.99
9.81 - 10.00	.50	19.81 - 20.00	1.00	29.81 - 30.00	1.50	39.81 - 40.00	2.00	49.81 - 50.00	2.50	59.81 - 60.00	3.00

Tax begins at **20¢ ON NON-FOOD ITEMS**
Tax begins at **$1.00 ON MEALS CONSUMED ON PREMISES.**

When the total charge for meals on premises reaches $1.00 or more, combine the charge with all other taxable items to find the total taxable sale.

Name _____

Class _____ Date _____

TEST FOR CHAPTER 12, Form A

Angle A	Tan a	Angle a	Tan a
10	0.18	50	1.19
20	0.36	60	1.73
30	0.58	70	2.75
40	0.84	80	5.67
45	1.00	89	57.29

1) Use the table to compute the distance C in the triangle.

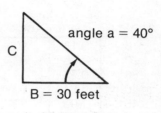

angle a = 40°

C

B = 30 feet

C = _____

2) Compute the square root of 235 to the nearer tenth. Use the "divide and average" method.

_____

3) Mr. Sparks measured 1.5 amps on his electric blanket. If the voltage on the circuit was 120 volts, what was the resistance in ohms, to the nearer tenth of an ohm?

_____

4) Use common denominators for these fractions. Then arrange them from smallest to largest. Circle the next fraction smaller than $\frac{5}{8}$.

$$\frac{5}{8} \qquad \frac{3}{4} \qquad \frac{9}{16} \qquad \frac{3}{8}$$

_____

5) Measure this line segment to the nearer $\frac{1}{4}$ inch.

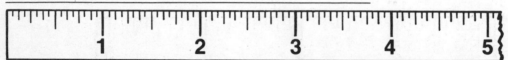

_____

6) Maurice is making a $\frac{3}{10}$ scale drawing of a doll house. How long should the line on the drawing be if it represents 20 inches?

_____

7) If a Driver Gear has 10 Teeth at 35 RPM, compute the number of teeth needed for the driven gear at 70 rpm.

_____

8) Louise bought $10.52 worth of food and paid for it with a 20-dollar bill. List her change in the least number of coins and bills.

Amount of Sale	Tax	Amount of Sale	Tax
16.01 - 16.20	.81	26.01 - 26.20	1.31
16.21 - 16.40	.82	26.21 - 26.40	1.32
16.41 - 16.60	.83	26.41 - 26.60	1.33
16.61 - 16.80	.84	26.61 - 26.80	1.34
16.81 - 17.00	.85	26.81 - 27.00	1.35
17.01 - 17.20	.86	27.01 - 27.20	1.36
17.21 - 17.40	.87	27.21 - 27.40	1.37
17.41 - 17.60	.88	27.41 - 27.60	1.38
17.61 - 17.80	.89	27.61 - 27.80	1.39
17.81 - 18.00	.90	27.81 - 28.00	1.40
18.01 - 18.20	.91	28.01 - 28.20	1.41
18.21 - 18.40	.92	28.21 - 28.40	1.42
18.41 - 18.60	.93	28.41 - 28.60	1.43
18.61 - 18.80	.94	28.61 - 28.80	1.44
18.81 - 19.00	.95	28.81 - 29.00	1.45
19.01 - 19.20	.96	29.01 - 29.20	1.46

_____

_____

9) Dominic paid $17.86 for a table lamp. Compute the total cost, including a 5% sales tax.

_____

10) Jerry bought a rug for $23.67 and a dish for $3.50. Find the total cost, including a 5% sales tax.

_____

Name _____

Class _____ Date _____

TEST FOR CHAPTER 12, Form B

Angle a	Tan a	Angle a	Tan a
10	0.18	50	1.19
20	0.36	60	1.73
30	0.58	70	2.75
40	0.84	80	5.67
45	1.00	89	57.29

1) Use the table to compute the distance C in the triangle.

angle a = 30°

C

B = 20 feet

C = _____

2) Compute the square root of 532 to the nearer tenth. Use the "divide and average" method.

_____

3) Mr. Sparks measured 1.2 amps on his electric blanket. If the voltage on the circuit was 120 volts, what was the resistance in ohms, to the nearer tenth of an ohm?

_____

4) Use common denominators for these fractions. Then arrange them from smallest to largest. Circle the next fraction smaller than $\frac{5}{8}$.

$$\frac{9}{16} \qquad \frac{3}{8} \qquad \frac{1}{4} \qquad \frac{5}{8}$$

_____

5) Measure this line segment to the nearer $\frac{1}{4}$ inch.

_____

6) Patrice is making a $\frac{3}{10}$ scale drawing of a doll house. How long should a line on the drawing be if it represents 15 inches?

_____

7) If a Driver Gear has 10 Teeth at 35 RPM, compute the number of teeth needed for the driven gear at 90 rpm.

_____

8) Sarah bought $11.23 worth of food and paid for it with a 20-dollar bill. List her change in the least number of coins and bills.

Amount of Sale	Tax	Amount of Sale	Tax
16.01 - 16.20	.81	26.01 - 26.20	1.31
16.21 - 16.40	.82	26.21 - 26.40	1.32
16.41 - 16.60	.83	26.41 - 26.60	1.33
16.61 - 16.80	.84	26.61 - 26.80	1.34
16.81 - 17.00	.85	26.81 - 27.00	1.35
17.01 - 17.20	.86	27.01 - 27.20	1.36
17.21 - 17.40	.87	27.21 - 27.40	1.37
17.41 - 17.60	.88	27.41 - 27.60	1.38
17.61 - 17.80	.89	27.61 - 27.80	1.39
17.81 - 18.00	.90	27.81 - 28.00	1.40
18.01 - 18.20	.91	28.01 - 28.20	1.41
18.21 - 18.40	.92	28.21 - 28.40	1.42
18.41 - 18.60	.93	28.41 - 28.60	1.43
18.61 - 18.80	.94	28.61 - 28.80	1.44
18.81 - 19.00	.95	28.81 - 29.00	1.45
19.01 - 19.20	.96	29.01 - 29.20	1.46

_____

_____

9) Becky paid $16.95 for a table lamp. Compute the total cost, including a 5% sales tax.

_____

10) Robert bought a rug for $26.81 and a dish for $1.94. Find the total cost, including a 5% sales tax.

_____

ANSWER KEY FOR
ACTIVITY WORKSHEETS AND CHAPTER TESTS

Survey Test for Whole Numbers

1) Tens
2) Hundreds
3) Thousands
4) Millions
5) Fifty-two thousand, six hundred nine
6) Two million, five hundred eighty-two thousand, eight hundred forty-four.
7) 470
8) 2,475,500
9) 0
10) 6023
11) 8908
12) 7308
13) 10240
14) 2071
15) 40794
16) $1510 \frac{1}{5}$
17) $102 \frac{37}{46}$
18) 2003
19) 16
20) 10

Whole Number Practice 1

1) 577
2) 182
3) 230
4) 1,160
5) 1,243
6) 1,177
7) 4,011
8) 5,286
9) 123,009
10) 106,762
11) 11,542
12) 6,939
13) 41,008
14) 527,142
15) 169,573
16) 129,056
17) 1,765,696
18) 49,991,528

Whole Number Practice 2

1) 327
2) 168
3) 3,293
4) 8,518
5) 3,486
6) 148
7) 550
8) 11,996
9) 24,483
10) 74,878
11) 46,016
12) 80,007
13) 451,208
14) 897
15) 72,398
16) 328,891
17) 247,217
18) 107,981
19) 251,187
20) 79,406
21) 61,139
22) 849,557
23) 109,398
24) 752,779
25) 753,713
26) 495,576
27) 12,162
28) 331,744

Whole Number Practice 3

1) 1,912
2) 2,436
3) 15,035
4) 3,839
5) 57,816
6) 33,768
7) 79,618
8) 22,325
9) 24,735
10) 55,836
11) 131,004
12) 823,550
13) 302,022
14) 1,785,978
15) 471,656
16) 988,325
17) 2,271,184
18) 1,156,608

Whole Number Practice 4

1) 271
2) 452
3) 206
4) 391
5) 623
6) 523
7) 305
8) 106
9) 615
10) 710
11) 564
12) 2,210
13) 107
14) 462
15) 107

Whole Number Practice 5

1) $293 \frac{1}{8}$
2) $508 \frac{3}{7}$
3) $350 \frac{5}{11}$
4) $666 \frac{5}{9}$
5) $309 \frac{1}{6}$
6) $363 \frac{1}{31}$
7) $339 \frac{6}{11}$
8) $5,500 \frac{19}{20}$
9) $117 \frac{3}{41}$
10) $107 \frac{5}{42}$
11) $418 \frac{1}{61}$
12) $360 \frac{3}{8}$
13) $159 \frac{3}{91}$
14) $1,047 \frac{11}{51}$
15) $1,114 \frac{2}{13}$

Review of Basic Operations/Whole Numbers

1) 366
2) 6,992
3) 618
4) $1,050 \frac{11}{25}$
5) 317,100
6) 77,411
7) $103 \frac{9}{34}$
8) 902,048
9) 56,410
10) $1,494 \frac{20}{33}$
11) 39,896
12) 31
13) $1,140 \frac{1}{4}$
14) 1,142
15) 7,505
16) 1,602
17) 10,741
18) 16,542
19) 264
20) 14,262
21) $13,530 \frac{1}{3}$
22) 2,360,000
23) 2,067,800
24) $1,038 \frac{31}{39}$
25) 7,793
26) $870 \frac{21}{65}$
27) 1,365
28) 18,130
29) 49,566
30) 5,664
31) 395
32) 4,004
33) 50,640
34) 50,837,787
35) 78,899
36) $704 \frac{2}{3}$
37) 10,045
38) 1,052
39) 900
40) 323
41) $1,035 \frac{7}{29}$
42) 2,534,058
43) $4,833 \frac{11}{12}$
44) 41,837
45) 85,157
46) $60,435 \frac{6}{7}$
47) 7,308
48) $12 \frac{438}{2501}$

Survey Test for Decimals

1) Thousandths
2) Ten-thousandths
3) Hundredths
4) Tenths
5) Thirty-four and seventy-two thou-sandths
6) Ten thousand, eight hundred fifty-three hundred-thousandths
7) 4.0
8) 46.15
9) 0.09
10) 183.7291
11) 21.969
12) 5.712
13) 24.2342
14) 56.353
15) 1.85
16) 0.53
17) 0.063
18) 0.9
19) 0.14
20) 0.667
21) 0.0000279
22) 3.5245
23) 519.87
24) 0.0000506

Decimal Practice 1

1) 6.71
2) 35.0532
3) 97.471
4) 69.5931
5) 66.5721
6) 18.867
7) 111.6093
8) 96.2692
9) 124.801
10) 1,003.86002
11) 2,996.39311
12) 490.1005
13) 94.068
14) 12.63
15) 16.9
16) 8.06
17) 7.206
18) 79.12
19) 40.802
20) 73.36

Decimal Practice 2

1) 28.66
2) 3.651
3) 6.472
4) 3.896
5) 44.731
6) 39.3629
7) 6.5995
8) 348.948
9) 1.5522
10) 58.0183
11) 37.9727
12) 8.1
13) 2.9766
14) 69.398
15) 7,464.2902
16) 28.606
17) 18.84
18) 3.52
19) 58.691
20) 14.66
21) 79.698
22) 31.1009
23) 5.5949
24) 0.97277

Decimal Practice 3

1) 8.84
2) 20.822
3) 0.3624
4) 15.998
5) 18.468
6) 10.3523
7) 209.502
8) 0.001872
9) 0.00927
10) 0.0000703
11) 0.000026796
12) 0.0002016
13) 9.153
14) 0.0441
15) 0.00096
16) 168.3
17) 0.010215
18) 0.00007429
19) 21.5754
20) 0.0007839

Decimal Practice 4

1) 4.6
2) 0.73
3) 5.2
4) 2.7
5) 4.6
6) 77
7) 1.7
8) 1.91
9) 52
10) 1.62
11) 19
12) 11.3
13) 0.421
14) 1.23
15) 91
16) 21.1
17) 1.82
18) 41.2
19) 0.137
20) 1.17
21) 2.7
22) 0.112

Decimal Practice 5

1) 4.01
2) 0.062
3) 0.023
4) 2.06
5) 1.07
6) 0.013
7) 5.01
8) 0.028
9) 110.1
10) 12.01
11) 0.039
12) 6.06
13) 32.01
14) 0.0091
15) 1.06
16) 12.01
17) 1.001
18) 2.01
19) 80.1
20) 11.03
21) 1.101
22) 9.01
23) 0.011
24) .202

Decimal Practice 6

1) 8.3
2) 118.33
3) 0.04
4) 1.017
5) 1.32
6) 1.0
7) 0.11
8) 8.55
9) 66.667
10) 0.10
11) 1.41
12) 43
13) 0.89
14) 1.1
15) 0.333
16) 0.12
17) 0.67
18) 10.63
19) 0.63
20) 15.63

Review of Basic Operations/Decimals

1) 8.241
2) 3.569
3) 1.395
4) 5.05
5) 6.1988
6) .024768
7) 10203.4
25) 385.067
26) 9.71
27) 2.21
28) 86.239
29) 7.11
30) 91.28
31) 2.04

8) 2.09
9) 2.346
10) 97.36
11) 5.11
12) 5.2
13) .0782
14) 36.584
15) 7.08
16) 766.559
17) 5.9
18) .000162
19) .060903
20) 93.703
21) .00899
22) 2.754
23) .12423
24) 385.067

32) 8.01
33) 2.58
34) .01244
35) .598
36) 431.83
37) 60.72
38) .0287
39) 45.79
40) 62.32
41) 139.92
42) 1.091
43) .008
44) .07926
45) 82.456
46) .8972
47) 513.848
48) 1.0201

33) $\frac{35}{112}$ 34) $\frac{8}{76}$ 35) $\frac{35}{91}$ 36) $\frac{42}{105}$

37) $\frac{36}{117}$ 38) $\frac{77}{161}$ 39) $\frac{175}{250}$ 40) $\frac{25}{200}$

Survey Test for Fractions

1) < 2) > 3) >

4) $\frac{3}{8}$ 5) $\frac{1}{3}$ 6) $14\frac{3}{11}$ 7) $\frac{8}{15}$

8) $\frac{44}{7}$ 9) $\frac{32}{9}$ 10) $\frac{105}{8}$ 11) $\frac{55}{12}$

12) $3\frac{1}{11}$ 13) 7 14) $21\frac{1}{4}$ 15) $7\frac{2}{3}$

16) $\frac{2}{63}$ 17) $3\frac{19}{24}$ 18) $22\frac{2}{9}$

19) $1\frac{5}{9}$ 20) $1\frac{1}{10}$ 21) $\frac{1}{26}$

22) $8\frac{1}{5}$ 23) $37\frac{7}{24}$ 24) $39\frac{4}{5}$

25) $10\frac{3}{8}$ 26) $33\frac{5}{14}$

Practice with Fractions, 1

1) $\frac{35}{40}$ 2) $\frac{16}{36}$ 3) $\frac{8}{12}$ 4) $\frac{25}{55}$

5) $\frac{15}{36}$ 6) $\frac{10}{35}$ 7) $\frac{36}{54}$ 8) $\frac{5}{10}$

9) $\frac{15}{39}$ 10) $\frac{20}{75}$ 11) $\frac{18}{66}$ 12) $\frac{4}{34}$

13) $\frac{36}{60}$ 14) $\frac{55}{60}$ 15) $\frac{16}{84}$ 16) $\frac{3}{48}$

17) $\frac{15}{65}$ 18) $\frac{20}{110}$ 19) $\frac{40}{56}$ 20) $\frac{57}{95}$

21) $\frac{18}{54}$ 22) $\frac{9}{63}$ 23) $\frac{72}{108}$ 24) $\frac{39}{52}$

25) $\frac{72}{126}$ 26) $\frac{22}{121}$ 27) $\frac{15}{80}$ 28) $\frac{64}{80}$

29) $\frac{14}{84}$ 30) $\frac{50}{70}$ 31) $\frac{12}{72}$ 32) $\frac{9}{54}$

Practice with Fractions, 2

1) $\frac{4}{5}$ 2) $10\frac{5}{6}$ 3) $\frac{1}{3}$ 4) $\frac{8}{9}$

5) $\frac{1}{2}$ 6) $\frac{5}{8}$ 7) $3\frac{1}{5}$ 8) $8\frac{7}{11}$

9) $\frac{14}{19}$ 10) $\frac{13}{18}$ 11) $\frac{13}{16}$ 12) $\frac{1}{5}$

13) $\frac{3}{7}$ 14) $2\frac{10}{21}$ 15) $\frac{2}{3}$ 16) $\frac{5}{7}$

17) $\frac{14}{19}$ 18) $6\frac{2}{5}$ 19) $17\frac{5}{9}$ 20) $11\frac{1}{5}$

21) $\frac{2}{9}$ 22) $\frac{1}{3}$ 23) $\frac{5}{7}$ 24) $\frac{3}{7}$

25) $\frac{7}{9}$ 26) $\frac{4}{5}$ 27) $\frac{7}{9}$ 28) $9\frac{3}{5}$

29) $\frac{2}{3}$ 30) $\frac{8}{9}$ 31) $\frac{5}{17}$ 32) $\frac{15}{19}$

33) $\frac{2}{7}$ 34) $\frac{2}{3}$ 35) $\frac{27}{34}$ 36) $\frac{9}{13}$

37) $\frac{5}{7}$ 38) $\frac{3}{7}$ 39) $7\frac{2}{3}$ 40) $23\frac{19}{21}$

Practice with Fractions, 3

1) $3\frac{3}{5}$ 2) $17\frac{1}{3}$ 3) $9\frac{1}{2}$

4) $3\frac{1}{7}$ 5) $8\frac{1}{3}$ 6) $5\frac{3}{5}$

7) $4\frac{3}{5}$ 8) $5\frac{1}{2}$ 9) $24\frac{7}{9}$

10) $3\frac{1}{6}$ 11) $8\frac{2}{5}$ 12) $4\frac{3}{8}$

13) 2 14) $4\frac{4}{7}$ 15) $26\frac{1}{4}$

16) $3\frac{3}{10}$ 17) $15\frac{1}{2}$ 18) $4\frac{1}{7}$

19) $9\frac{1}{2}$ 20) $9\frac{1}{7}$ 21) 12

22) $11\frac{2}{11}$ 23) $6\frac{3}{7}$ 24) $38\frac{1}{3}$

25) 7 26) $5\frac{4}{7}$ 27) $5\frac{1}{3}$

28) $5\frac{7}{8}$ 29) $4\frac{1}{13}$ 30) $3\frac{1}{2}$

31) $5\frac{3}{10}$ 32) $9\frac{3}{8}$ 33) $7\frac{1}{4}$

34) $8\frac{1}{3}$ 35) $14\frac{1}{4}$

Practice with Fractions, 4

1) $15\frac{5}{8}$ 2) $28\frac{7}{17}$

3) $27\frac{7}{10}$ 4) $11\frac{7}{26}$

5) $5\frac{15}{56}$ 6) $8\frac{7}{12}$

7) $14\frac{2}{3}$ 8) $1\frac{1}{30}$

9) $\frac{13}{22}$ 10) $3\frac{11}{20}$

11) $39\frac{55}{56}$ 12) $17\frac{2}{5}$

13) $33\frac{19}{24}$ 14) $\frac{4}{15}$

15) $1\frac{5}{14}$ 16) $5\frac{13}{24}$

17) $7\frac{14}{15}$ 18) $11\frac{5}{18}$

19) $13\frac{17}{66}$ 20) $2\frac{17}{45}$

21) $10\frac{15}{38}$ 22) $33\frac{7}{32}$

23) $7\frac{5}{13}$ 24) $26\frac{13}{28}$

Practice with Fractions, 6

1) $8\frac{12}{13}$ 2) $1\frac{5}{7}$

3) $10\frac{11}{12}$ 4) $15\frac{12}{35}$

5) $32\frac{17}{18}$ 6) $37\frac{9}{10}$

7) $31\frac{5}{14}$ 8) $42\frac{73}{90}$

9) $1\frac{14}{15}$ 10) $24\frac{9}{16}$

11) $7\frac{9}{11}$ 12) $24\frac{53}{62}$

13) $22\frac{11}{12}$ 14) $1\frac{19}{24}$

15) $\frac{8}{15}$ 16) $\frac{71}{72}$

17) $24\frac{7}{16}$ 18) $1\frac{5}{6}$

19) $15\frac{4}{5}$ 20) $13\frac{13}{18}$

21) $5\frac{3}{4}$ 22) $9\frac{17}{20}$

23) $\frac{8}{9}$ 24) $1\frac{6}{7}$

Practice with Fractions 5

1) $\frac{2}{7}$ 2) $12\frac{2}{3}$

3) $2\frac{1}{2}$ 4) $5\frac{1}{5}$

5) $2\frac{3}{4}$ 6) $23\frac{19}{36}$

7) $1\frac{11}{21}$ 8) $9\frac{5}{32}$

9) $3\frac{1}{3}$ 10) $9\frac{4}{5}$

11) $22\frac{1}{24}$ 12) $1\frac{6}{11}$

13) $1\frac{7}{16}$ 14) $6\frac{11}{36}$

15) $15\frac{1}{3}$ 16) $5\frac{13}{18}$

17) $24\frac{17}{24}$ 18) $5\frac{7}{36}$

19) $1\frac{15}{28}$ 20) $9\frac{3}{40}$

21) $26\frac{5}{39}$ 22) $13\frac{1}{14}$

23) $78\frac{1}{11}$ 24) $9\frac{13}{25}$

Practice with Fractions, 7

1) $\frac{11}{4}$ 2) $\frac{3}{2}$ 3) $\frac{4}{3}$ 4) $\frac{21}{8}$

5) $\frac{11}{3}$ 6) $\frac{31}{6}$ 7) $\frac{22}{7}$ 8) $\frac{14}{3}$

9) $\frac{32}{5}$ 10) $\frac{26}{5}$ 11) $\frac{29}{6}$ 12) $\frac{22}{5}$

13) $\frac{44}{7}$ 14) $\frac{19}{5}$ 15) $\frac{85}{9}$ 16) $\frac{51}{7}$

17) $\frac{23}{9}$ 18) $\frac{65}{8}$ 19) $\frac{21}{2}$ 20) $\frac{35}{3}$

21) $\frac{28}{3}$ 22) $\frac{50}{3}$ 23) $\frac{47}{4}$ 24) $\frac{41}{4}$

25) $\frac{92}{11}$ 26) $\frac{57}{10}$ 27) $\frac{77}{3}$ 28) $\frac{62}{3}$

29) $\frac{23}{2}$ 30) $\frac{37}{2}$ 31) $\frac{53}{2}$ 32) $\frac{37}{2}$

33) $\frac{417}{20}$ 34) $\frac{101}{12}$ 35) $\frac{57}{11}$ 36) $\frac{142}{4}$

37) $\frac{47}{3}$ 38) $\frac{131}{4}$ 39) $\frac{9}{5}$ 40) $\frac{10}{7}$

41) $\frac{35}{6}$ 42) $\frac{40}{3}$ 43) $\frac{149}{12}$ 44) $\frac{41}{5}$

45) $\frac{225}{11}$ 46) $\frac{79}{5}$ 47) $\frac{41}{3}$ 48) $\frac{89}{10}$

49) $\frac{35}{2}$ 50) $\frac{91}{4}$ 51) $\frac{61}{5}$ 52) $\frac{83}{8}$

Practice with Fractions, 8

1) $\frac{5}{9}$ 2) $\frac{3}{5}$ 3) $\frac{5}{16}$ 4) $\frac{5}{16}$

5) $\frac{2}{13}$ 6) $\frac{1}{4}$ 7) $\frac{5}{26}$ 8) $\frac{1}{4}$

9) 2 10) $2\frac{1}{5}$ 11) $\frac{13}{14}$ 12) $\frac{5}{9}$

13) $3\frac{1}{5}$ 14) $2\frac{13}{18}$ 15) $3\frac{12}{35}$ 16) 3

17) $3\frac{4}{5}$ 18) $\frac{2}{7}$ 19) $\frac{37}{49}$ 20) 10

21) $5\frac{1}{7}$ 22) $3\frac{1}{3}$ 23) 21 24) 23

25) $26\frac{1}{4}$ 26) $7\frac{1}{3}$ 27) $2\frac{1}{10}$ 28) $20\frac{1}{3}$

Practice with Fractions, 9

1) $\frac{6}{7}$ 2) 3 3) 1

4) $\frac{5}{6}$ 5) $\frac{7}{8}$ 6) $\frac{15}{26}$

7) $1\frac{1}{15}$ 8) $1\frac{1}{4}$ 9) 1

10) $3\frac{1}{3}$ 11) $2\frac{4}{15}$ 12) $1\frac{1}{11}$

13) $7\frac{1}{3}$ 14) $1\frac{9}{13}$ 15) 1

16) $1\frac{1}{2}$ 17) $1\frac{1}{14}$ 18) $2\frac{22}{35}$

19) $\frac{35}{136}$ 20) 12 21) 4

22) $14\frac{2}{3}$ 23) 12 24) $37\frac{1}{3}$

Review of Basic Operations/Fractions

1) $\frac{5}{14}$ 17) $\frac{53}{56}$ 33) $\frac{19}{24}$

2) $5\frac{38}{45}$ 18) $2\frac{2}{7}$ 34) $8\frac{12}{23}$

3) $13\frac{13}{24}$ 19) $1\frac{2}{13}$ 35) $31\frac{7}{72}$

4) $6\frac{1}{24}$ 20) 2 36) $10\frac{47}{50}$

5) 1 21) 2 37) $13\frac{1}{42}$

6) $10\frac{1}{9}$ 22) $\frac{11}{20}$ 38) $\frac{21}{34}$

7) $5\frac{2}{7}$ 23) $9\frac{6}{7}$ 39) $27\frac{43}{81}$

8) $\frac{27}{50}$ 24) $18\frac{4}{15}$ 40) $21\frac{12}{13}$

9) $64\frac{1}{25}$ 25) $15\frac{23}{24}$ 41) $\frac{1}{5}$

10) $\frac{1}{5}$ 26) $1\frac{21}{22}$ 42) $\frac{2}{13}$

11) $2\frac{3}{5}$ 27) $15\frac{11}{18}$ 43) $\frac{36}{65}$

12) $2\frac{8}{15}$ 28) $34\frac{37}{39}$ 44) 87

13) $6\frac{4}{11}$ 29) $\frac{1}{15}$ 45) $\frac{9}{14}$

14) $4\frac{12}{17}$ 30) $16\frac{1}{5}$ 46) 1

15) $14\frac{1}{4}$ 31) $12\frac{9}{28}$ 47) $69\frac{5}{9}$

16) $19\frac{14}{15}$ 32) $\frac{1}{30}$ 48) $52\frac{12}{35}$

Review of Basic Operations

1) 2,246 17) 2.2658 33) $\frac{1}{2}$

2) 0.009 18) 578 34) 12.69

3) $\frac{34}{63}$ 19) 820,000 35) 12.5

4) 2.547 20) $\frac{3}{4}$ 36) 0.0006

5) 12,721 21) $11\frac{7}{17}$ 37) $78\frac{3}{4}$

6) 52.52 22) $3\frac{19}{24}$ 38) $32\frac{17}{22}$

7) 12.331 23) 1.38 39) $1,060\frac{1}{2}$

8) $\frac{1}{4}$ 24) $27\frac{13}{27}$ 40) 108

9) 1,237 25) $3\frac{1}{3}$ 41) $\frac{1}{3}$

10) 6.09203 26) 1,228 42) $1,008\frac{11}{60}$

11) 21.12 27) $\frac{13}{22}$ 43) 936.93

12) 0.117 28) $17\frac{1}{9}$ 44) 95.54

13) 1.242 29) 208 45) 55.1

14) $1\frac{59}{121}$ 30) 10,100 46) $8\frac{22}{25}$

15) $12\frac{5}{6}$ 31) $21\frac{4}{11}$ 47) 405.24

16) 0.0001 32) 47,018 48) 0.002002

Test for Chapter 1, Form A

1) $278.13 2) 9 hrs., 30 min.
3) $59.85 4) $19.62
5) $129.70 6) $5.61
7) 0.05 8) 0.17
9) 1.1 10) $190.20
11) $81.00 12) $544.37
13) $250.76

Test for Chapter, 1 Form B

1) $333.50 2) 9 hours
3) $62.72 4) $18.94
5) $138.03 6) $2.51
7) 0.06 8) 0.19
9) .005 10) $280.60
11) $89.60 12) $551.37
13) $245.75

Test for Chapter 2, Form A
1) 465¢   2) 72¢   3) 12¢
4) $0.53   5) $1.95   6) $0.03
7) 45¢
8) $4.83
9) $15.17
10) $0.0175
11) $0.021
12) $0.11
13) $0.1227

Test for Chapter 2, Form B
1) 345¢   2) 45¢   3) 3¢
4) $0.46   5) $1.93   6) $0.04
7) 34¢
8) $4.45
9) $15.55
10) $0.016
11) $0.017
12) $0.099
13) $0.112

Test for Chapter 3, Form A
1) $20.82
2) $4.11
3) $7.60
4) $36.76
5) 22% discount
6) $8.99
7) $3.04
8) $43.68
9) $116.73
10) $100.00

Test for Chapter 3, Form B
1) $20.09
2) $4.34
3) $24.90
4) $30.47
5) 19% discount
6) $7.85
7) $3.46
8) $44.85
9) $116.73
10) $112.50

Meters to Read
1) 2219   2) 2619
3) 4776   4) 5823
5) 0246   6) 1771

7) 6143   8) 6464
9) 1831   10) 5912
11) 2752   12) 4692
13) 7145   14) 3712
15) 8610   16) 1717

Test for Chapter 4, Form A
1) $283.00
2) $30,000.00
3) $9600.00 down payment
   $38,400.00 to be financed
4) 7614 units   5) 8322 units
6) 708 units used
7) $18.56 total charge
8) $109.05 total charge for the month
9) $255.60 annual payment
10) $278.80 annual payment

Test for Chapter 4, Form B
1) $294.00
2) $37,500
3) $9,900.00 down payment
   $39,600 to be financed
4) 7611 units   5) 8136 units
6) 525 units used
7) $17.32 total charge
8) $104.20 total charge for the month
9) $257.14 annual payment
10) $302.40 annual payment

Test for Chapter 5, Form A
1) $6322.00 total price
2) $3800.00 actual cost
3) $140.00 total premium
4) 18.8 mpg
5) $24.50
6) 480 miles
7) 55 mph
8) 8 hours
9) 23 gallons
10) $94.89 total cost

Test for Chapter 5, Form B
1) $6362.00 total price
2) $3804.00 actual cost
3) $148.00 total premium
4) 18.5 mpg
5) $27.00
6) 675 miles
7) 47.5 mph

8) 9 hours
9) 23.3 gallons
10) $91.96 total cost

Test for Chapter 6, Form A

1) 590 calories
2) 1860 calories
3) 96 calories, rounded
4) 135 calories
5) .11%
6) 500 calories
7) 35 hours
8) 8 ounces, rounded
9) 3:20
10) a. 2:25
    b. 5:20

Test for Chapter 6, Form B

1) 900 calories
2) 615 calories
3) 122, rounded
4) 108 calories
5) 5%
6) 750 calories
7) 28 at 125 cal./hr.
   29 at 120 cal./hr.
8) $6\frac{2}{3}$ ounces
9) 1:35
10) a. 3:35   b. 5:10

Test for Chapter 7, Form A

1) $455.00 original price
2) 20% rate of discount
3) April 24
4) Perimeter = 20 in.
   Area = 16 sq. in.
5) 4 quarts
6) 320 sq. ft.
7) $241.80
8) 51 ft.
9) $10,140
10) 180 feet

Test for Chapter 7, Form B

1) $493.75
2) 32% rate of discount
3) April 22
4) Perimeter = 40 in.
   Area = 64 sq. in.

5) 4 quarts
6) 336 sq. ft.
7) $261.95
8) 55 feet
9) $10,290
10) 190 feet

Test for Chapter 8, Form A

1) 540 miles
2) 324 miles
3) 3 hr., 55 min.
4) $503.13
5) $51, rounded
6) 201 marks
7) $384.00
8) $9.50
9) 5:40 P.M.
10) 11:20 P.M.

Test for Chapter 8, Form B

1) 630 miles
2) 378 miles
3) 4 hrs., 40 min.
4) $402.50
5) $47, rounded
6) 302 marks, rounded
7) $432.00
8) $7.50
9) 4:40 P.M.
10) 12:20 A.M.

Test for Chapter 9, Form A

1) $786.07 average
2) $827.67 average
3) Housing = $195.75
   Food = $145.00
   Clothes = $65.25
   Gifts = $50.75
   Transportation = $94.25
4) Entertainment = $36.25
   Miscellaneous = $21.75
   Insurance = $36.25
   Health = $21.75
   Savings = $58.00
5) $91.50 for food
6) 9% for food
7) Records, 7%
   Savings 29%
   Entertainment, 14%
   Hair, 17%
8) $53.25 for clothes

9) Food, 90°
Housing, 54°
Clothing, 72°
Car, 36°
Health, 36°
Miscellaneous, 72°

10)

Test for Chapter 9, Form B
1) $720.42 average
2) $925.17 average
3) Housing = $173.34
Food = $128.40
Clothes = $57.78
Gifts = $44.94
Transportation = $83.46
4) Entertainment = $32.10
Miscellaneous = $19.26
Insurance = $32.10
Health = $19.26
Savings = $51.36
5) $112.35 for food
6) 8% for food, rounded
7) Records, 5.4%
Savings, 27%
Entertainment, 13.5%
Hair, 16.2%
8) $55.25 for clothes
9) Food, 90°
Housing, 54°
Clothing, 72°
Car, 18°
Health, 54°
Miscellaneous, 72°

10)

Test for Chapter 10, Form A
1) $23.33 interest
2) $796.00 balance

3) 7.2 years
4) Three hundred forty-five and 72/100
One thousand twenty-two and 00/100
5) 1. $732.25
2. $629.74
3. $981.74
4. $731.74
5. $653.74
6) No, $0.50 difference
7) $3.625 or $3.63
8) a) $6.625 (or $6.63) Loss
b) $4.125 (or $4.13) Profit
9) 19% (rounded)
10) 6 shares (rounded)

Test for Chapter 10, Form B
1) $18.67 interest
2) $902.16 balance
3) 7.2 years
4) Two hundred thirty-five and 72/100
One thousand thirty-two and 00/100
5) 1. $730.66
2. $628.15
3. $973.15
4. $723.15
5. $645.15
6) No, $0.50 difference
7) $4.625 or $4.63
8) a) $3.375 profit   9) 27.4%
b) $4.875 loss   10) 6 shares

Federal Income Tax
Answers depend on current tax regulations.

Test for Chapter 11, Form A
1) 352,635,000,000
2) Two hundred sixty-five billion, four hundred one million
3) a. $3,848
b. $3,001
4) Amount to be refunded is $119.
5) 1.203, 1.23, 1.52
1.07, 10.06, 11.1
5.308, 5.32, 5.4
.701, 7.003, 7.02
.1, .101, 1.1
6) $24,360 assessed value
7) 2.95%
8) $842.14 tax
9) 1.6% effective tax rate
10) 1.2% effective tax rate
$624 with effective tax rate method

Test for Chapter 11, Form B
1) 352,642,000,000
2) Five hundred twenty-one billion, three hundred six million
3) a. $4,167
   b. $3,280
4) The refund is $899.
5) 1.09, 1.32, 1.41
   1.99, 10.19, 10.9
   5.308, 5.33, 5.4
   .600, 6.02, 60.1
   .2, .202, 2.2
6) $20,640 assessed value
7) 2.8%
8) $616.85 tax
9) 1.2% effective tax rate
10) 1.3% effective tax rate
    $663 property tax

Test for Chapter 12, Form A
1) 25.2 feet
2) 15.3
3) 80 ohms
4) $\frac{6}{16}$ $\boxed{\frac{9}{16}}$ $\frac{10}{16}$ $\frac{12}{16}$
5) $3\frac{3}{4}$ in.

6) 6 in.
7) 5 teeth
8) 3 pennies
   2 dimes
   1 quarter
   4 one dollar bills
   1 five dollar bill
9) $0.90 sales tax, $18.76 total
10) $1.36 sales tax, $28.53 total

Test for Chapter 12, Form B
1) 11.6 feet
2) 23.1
3) 100 ohms
4) $\frac{4}{16}$ $\frac{6}{16}$ $\boxed{\frac{9}{16}}$ $\frac{10}{16}$
5) $3\frac{2}{4}$ in
6) $4\frac{1}{2}$ in.
7) 5 teeth
8) 2 pennies
   1 quarter
   1 half dollar
   3 one dollar bills
   1 five dollar bill
9) $0.85 sales tax, $17.80 total
10) $1.44 sales tax, $30.19 total

MEDIA MATERIALS, INC.
2936 Remington Avenue   Baltimore, Maryland 21211

EASY-ORDER FORM
Toll-Free   1-800-638-1010
In MD   1-301-235-1700

HIGH INTEREST / EASY READING
Basic Textbooks for Secondary and Adult Students

BILL TO _____

School _____

Street _____

City _____ State _____ Zip _____

Ordered By _____ Title _____

☐ Check enclosed. SAVE: We pay all shipping and handling charges when payment accompanies your order.

☐ Bill me. Phone No. _____

SHIP TO _____

School _____

Street _____

City _____ State _____ Zip _____

Phone No. _____

☐ Charge   ☐ Master Card   ☐ Visa

Card No. _____ Expires _____

Signature _____

QTY.	ORDER NO.	TITLE	LEVEL	PRICE / EXTENSION PRICE

ENGLISH

QTY.	ORDER NO.	TITLE	LEVEL	PRICE	EXTENSION
_____	10050	**English to Use**		$12.95	_____
_____	10051	Manual/Resource Book	**7-12, Adult**	$9.95	_____
_____	10052	Student Workbook	*Reading level: 2.6*	$3.95	_____
_____	10053	Workbook Answer Key		$1.00	_____
_____	10054	35 Blackline Masters for Testing/Practice		$18.95	_____
_____	10010	**Basic English Grammar**		$12.95	_____
_____	10011	Manual/Resource Book	**7-12, Adult**	$9.95	_____
_____	10012	Student Workbook	*Reading level: 2.6*	$3.95	_____
_____	10013	Workbook Answer Key		$1.00	_____
_____	10014	14 Blackline Masters for Testing/Practice		$12.95	_____
_____	10030	**Basic English Composition**		$12.95	_____
_____	10031	Manual/Resource Book	**7-12, Adult**	$9.95	_____
_____	10032	Student Workbook	*Reading level: 2.8*	$3.95	_____
_____	10033	Workbook Answer Key		$1.00	_____
_____	10034	32 Blackline Masters for Testing/Practice		$18.95	_____
_____	10070	**Life Skills English**		$12.95	_____
_____	10071	Manual/Resource Book	**7-12, Adult**	$9.95	_____
_____	10072	Student Workbook	*Reading level: 2.8*	$3.95	_____
_____	10073	Workbook Answer Key		$1.00	_____
_____	10074	35 Blackline Masters for Testing/Practice		$18.95	_____
_____	10080	**English for Careers**		$12.95	_____
_____	10081	Manual/Resource Book	**7-12, Adult**	$9.95	_____
_____	10082	Student Workbook	*Reading level: 3.0*	$3.95	_____
_____	10083	Workbook Answer Key		$1.00	_____
_____	10084	32 Blackline Masters for Testing/Practice		$18.95	_____

MATHEMATICS

QTY.	ORDER NO.	TITLE	LEVEL	PRICE	EXTENSION
_____	10020	**Basic Mathematics Skills**		$12.95	_____
_____	10021	Manual/Resource Book	**7-12, Adult**	$9.95	_____
_____	10022	Student Workbook	*Reading level: 2.8*	$3.95	_____
_____	10023	Workbook Answer Key		$1.00	_____
_____	10024	50 Blackline Masters for Testing/Practice		$24.95	_____
_____	10040	**Life Skills Mathematics**		$12.95	_____
_____	10041	Manual/Resource Book	**7-12, Adult**	$9.95	_____
_____	10042	Student Workbook	*Reading level: 2.8*	$3.95	_____
_____	10043	Workbook Answer Key		$1.00	_____
_____	10044	68 Blackline Masters for Testing/Practice		$24.95	_____
_____	10060	**Mathematics for Consumers**		$12.95	_____
_____	10061	Manual/Resource Book	**7-12, Adult**	$9.95	_____
_____	10062	Student Workbook	*Reading level: 2.8*	$3.95	_____
_____	10063	Workbook Answer Key		$1.00	_____
_____	10064	56 Blackline Masters for Testing/Practice		$24.95	_____

Double Guarantee

SATISFACTION GUARANTEED on every order. If, for any reason, you are not satisfied with your purchase, just return it and we will replace it, refund the purchase price or credit your account, even after the merchandise has been used. If any cassette tape or diskette is damaged in normal classroom use, return it and a replacement will be sent at no charge.

SERVICE GUARANTEED on every order. We will ship your order complete within 3 to 5 days after receipt. Further, if you return any materials, your refund or exchange will be made promptly.

FOLD HERE AND TAPE CLOSED

FOLD HERE AND TAPE CLOSED